A Liberation for the Earth

A Liberation for the Earth

Race, Climate and Cross

A. M. Ranawana

© A. M. Ranawana 2022

Published in 2022 by SCM Press
Editorial office
3rd Floor, Invicta House,
108–114 Golden Lane,
London EC1Y 0TG, UK

www.scmpress.co.uk

SCM Press is an imprint of Hymns Ancient & Modern Ltd
(a registered charity)

Hymns Ancient & Modern® is a registered trademark of
Hymns Ancient & Modern Ltd
13A Hellesdon Park Road, Norwich,
Norfolk NR6 5DR, UK

All rights reserved. No part of this publication may be reproduced,
stored in a retrieval system, or transmitted,
in any form or by any means, electronic, mechanical,
photocopying or otherwise, without the prior permission of
the publisher, SCM Press.

The author has asserted her right under the Copyright, Designs and
Patents Act 1988 to be identified as the Author of this Work

Scripture quotations are from the Revised Standard Version of the
Bible—Second Catholic Edition (Ignatius Edition), copyright 2006
National Council of the Churches of Christ in the United States of
America. Used by permission. All rights reserved.

British Library Cataloguing in Publication data
A catalogue record for this book is available
from the British Library

ISBN 978-0-334-06126-7

Typeset by Regent Typesetting
Printed and bound in Great Britain by
CPI Group (UK) Ltd

Contents

Introduction   1

1   The Cry of the Earth: Understanding Ecological Sin and Ecological Conversion   16

2   All These Bleeding Wounds – Liberational Theologies and the Ecological Crisis   40

3   'The Plantation is Always With Us': The Histories and Presents of Extraction and Domination   66

4   A Rainbow Coalition: Faith-based Mobilizations in Northern and Southern Spaces   93

5   A Theology of Rage   118

Index   133

Introduction

Whence come famines and tornadoes and hailstorms, our present warning blow? ... And how is the creation, once ordered for the enjoyment of men, their common and equal delight, changed for the punishment of the ungodly, in order that we may be chastised through that for which, when honoured with it, we did not give thanks, and recognise in our sufferings that power which we did not recognise in our benefits?
Gregory of Nazianzus (*Oratio* 16.5)

Poetry can haunt you. The powerful words in a poem called 'Rise' written by Kathy Jetñil-Kijiner and Aka Niviâna connect two 'island' poets, one from the Marshall Islands and one from Kalaallit Nunaat (Greenland), bringing their realities of melting glaciers and rising sea levels.[1] The reality for them, as well as for me, a postulant theologian from a little island, is that our islands are sinking. I have been haunted by their words ever since I heard the poem, for the call echoes deep in my heart and my mind. I thought of these words in the summer of 2019, when I was taking a trip in a small boat in Negombo, Sri Lanka, going out to see the mangroves in the area. Coming upon the mangroves I drew back a little in shock. Covering the base of them was a brightly coloured plastic skirt of garbage. These ancient mangroves are legally protected due to their carbon-storing 'superpowers', and for the lives that thrive within their stilt-like roots – fish, prawns, crabs and other marine animals.[2] After the impact of the Boxing Day tsunami of 2004, Sri Lanka became the first country to protect all its mangrove forests.

A LIBERATION FOR THE EARTH

The sadness and emotion of this moment of seeing the plastic-covered mangroves, that chest-constricting thought of the island that I come from sinking under a wave of plastic and pollution, I later detailed in an essay for Hannah Malcolm's *Words for a Dying World*.

It is not only Sri Lanka that is sinking and dying. In July 2018, a nine-year-old girl was crushed and killed by a mine vehicle near the North Mara mining facility in Tanzania. The mine is owned by the London-listed Acacia Mining company, which in turn is mostly owned by Barrick, a Toronto-based firm. Acacia Mining has a long history of both environmental and human rights abuse, even being fined USD 2.4 million in 2019 for breaching Tanzanian environmental regulations. The waste from the mining facility releases heavy toxins into the air and water, causing irreparable environmental and public health damage. In response to attempts to investigate these crimes in the years prior to 2019, the company put in place its own security forces who often respond with excessive force. This is not only impacting investigators but also nearby villagers. As an article in *The Guardian* newspaper noted, the company had to admit to 32 'trespasser-related' fatalities between 2014 and 2017 alone.[3] Watchdogs also noted how the settler practices of the mining company severely disenfranchised the Indigenous Kuria community and their artisanal mining practices. Paula Butler summarizes it thus: Canadian-led resource extraction, still based on the idea of northward settler expansion, has in our modern world served only to exploit, displace and dispossess Indigenous populations across central America and Africa.[4]

Both these instances cause me grief and anger. And it is of anger that I write in this book. It is anger that flares when one considers how islands like Sri Lanka are at the 'receiving end' of historical and present colonization and racialized extraction. Of mining companies from the Global North that continue to extract and exploit land in Africa and South America, of large shipments of waste from Canada and the United Kingdom that are dumped in East and South Asia.

What these learnings did was underscore my thinking on

the need to further conversations on the topic of ecological sin and the importance of a spiritual awakening that would add to the ongoing activities that focus on effective transformative ecological justice. I was troubled also as to why, in the different discussions surrounding climate activism, there was limited conversation around colonialisms, both historical and present, and how the idea of some humans being seen as 'lesser' was fundamental to how the land was expropriated and extracted.

In 2016 I was working with Caritas within the Catholic community in Western Canada as a *personne animatrice*, teaching and embedding the values of Catholic social justice, especially the new encyclical *Laudato Si'*. Talking about ecological sin and ecological conversion presented me with a profound recognition of how much further our theologies needed to take us. Were we able to recognize our ecological sins and do penance for them? With this question in my head, I assisted in a research project that studied the reception of *Laudato Si'* among Christian communities in North America, and this process further underscored a variety of theological confrontations that needed to be made. Chief of these is that, before we can think about ecological sin or ecological conversion, we first need to bring back to the centre a justice narrative of how we, as Christians, approach our shared humanity. In 2021, then, I attempted to put these reflections into a book.

This book has been written in various notebooks, on post-its, in field notes, in conversations with activists and academics, in a workshop in a church in Calgary, Alberta, on a university campus in Dhaka, on long walks in rural England, by a ruined castle in Cruden Bay, Aberdeenshire, and on beaches in Sri Lanka. It is a book written from sadness and also from a place of rage, a generative, creative rage. By speaking of writing, I do not mean the physical act of putting words down on virtual paper, but the collecting of thoughts, readings, conversations, silences and experiences that formulate the arguments I sketch out in the book. What is written in this book is not my own original scholarship; instead, it is an attempt to bring together the work of activists, academics and theologians who are all endeavouring to argue for a project of multiple and complex justices.

In the initial planning for this writing, I had intended to use liberation theology as a framework to strengthen calls for an alliance between the fight for racial justice and the fight for ecological justice, and to speak from a space of lament. However, as the writing of the book went on, and as I considered where lament leads us and noticed the righteous anger of activists, academics and others in the Global South, I began to think about rage as a theological concept, and why such generative rage was fundamental for ensuring a justice narrative within eco-theologizing. I wonder how and why we must be affected by this emotional–theological category and allow it to move us towards a place of collective repentance, to engaging robustly with the concept of ecological sin and through this to being able to build what Keri Day calls 'beloved communities'.[5] By focusing on rage, the central aim of the book is to point to the voices of liberational and 'Third World' theologians and faith-based activists from the Global South and their insistant demands. A liberated Earth is one that has achieved the aims of the struggles against all injustices.

Why rage?

You may ask me, why rage? Indeed, when I gave an abstract of the argument of this book at a recent conference, this was the question that was immediately asked. Isn't rage a hostile position to take? Isn't it more important to sit with grief, to lament? My response is akin to that of Christina Sharpe: rage is important because 'some of us have never had any other choice'.[6]

Rage is a strong theme, not just among climate activists and academics from the Global South like me, but also in communities living in increasingly difficult situations even within 'Western' spaces. Nadine El-Enany's ongoing research, for example, points out the living conditions of immigrant and minority ethnic communities in the United Kingdom, detailing how such groups are often subject to state terror, austerity and poor housing and often being situated in the parts of cities affected by the worst air pollution. Being settled in spaces with extremely poor air quality leads to highly skewed health outcomes for racialized communities. This became a pointed case in the death of nine-year-old Ella Kissi Debrah, which called attention to the fact that Black and minority ethnic populations settled in inner-city areas are more frequently, and on a longer-term basis, exposed to air pollutants. It became even more evident during the time of the Covid pandemic when Black and minority ethnic sufferers from Covid-19 were found to be more vulnerable to the disease. As the Wretched of the Earth Collective and other aligned groups noted in an open letter to Extinction Rebellion,

> bleakness is not something of 'the future'. For those of us who are indigenous, working class, black, brown, queer, trans or disabled, the experience of structural violence became part of our birthright. Whenever the tide of ecological violence rises, our communities, especially in the Global South, are always first hit. We are the first to face poor air quality, hunger, public health crises, drought, floods and displacement.[7]

This is where the discussion of rage comes from. This is rage that has fuelled the struggle of communities for decades because the climate crisis is not a new crisis. This is rage that is connected to the argument that the struggle for ecological justice cannot be separated from the struggle to overturn systems of enslavement, expropriation, colonization and Indigenous genocide. Imperial projects that were centred on the expansion of empire also focused attention on the dominion over land such that entire landscapes were subjected to control and exploitation, and colonies were created in order to maximize extraction of natural resources. As Reyes-Carranza, writing on environmental justice in Brazil has noted, a disregard for Black and Indigenous lives in the Global South interweaves with other socio-political conditions, including lax environmental regulations and a lack of corporate accountability, helping to perpetuate climate and environmental injustice.[8] As the geographer Keston Perry has highlighted,[9] so much of the Caribbean, such as the Bahamas, Dominica and Antigua and Barbuda, has faced enormous losses from climate-induced natural disasters in recent times. This is a result of colonial systems that prioritized extractive plantation agriculture over protective ecosystems and disregarded Indigenous practices and knowledge about the environment.[10] This situation continues today, not in explicitly colonial ways, but through deforestation at local government levels, the presence of multinational corporations in Africa and South America, and the constant 'reclamation' of land for the purpose of limitless economic growth. Often this growth is at the expense of Indigenous and other marginalized communities. As Olufemi Taiwo, among others, has also argued, Global North countries are disproportionately responsible for the emissions that caused climate change, as well as the underdevelopment that made Africans and other colonized peoples susceptible to climate change and 'this destruction would in and of itself count as [ongoing] colonial violence'.[11] Hence, fighting for ecological justice is also part and parcel of anti-racist and anti-capitalist movements. It is an international struggle. The silencing and erasure of Indigenous communities and the exploitation and oppression of the vulnerable peoples from the Global South at climate talks must and does fuel the rage that demands an overturning of the 'death-creating' systems of capital.

Leon Sealey-Huggins argues that the problem lies in the ways in which the analysis of environmental problems gets framed in technical terms.[12] The only way to really understand the damage caused by climate change is by analysing it through and alongside broader social and political processes, including structural racism. Huggins provides the example of Haiti – one of the poorest countries in the Western hemisphere – which struggles with reparation payments to France. This keeps Haiti in debt and, when faced with an overwhelming climate disaster such as a hurricane, it hurtles more and more into debt and deprivation. Technical approaches fail, therefore, because they are reductionist, as well as performing a clinical separation between the politics of race and the politics of climate. The ability of a Canadian mining company to act in an extractive and exploitative manner in Tanzania or Chile is entirely because of embedded, structural racism. Y Ariadne Collins concurs, noting that due to the legacy of colonial and development encounters, people in the ethnically diverse Caribbean region are racialized through their continued and lived interaction with a natural environment to which they were forced to adapt.[13] Therefore a key component of the fight for climate justice is meaningful examination and study of the material and discursive relations through which climate harms are produced, and understanding who bears the burden of climate effects and how mitigation will impact upon us in different ways. Theologically, it requires dwelling on the moral imperative for transformative justice. This is especially underlined when we approach the Earth as the 'new poor'. We can ask ourselves what it means to understand the environment as a 'new poor'[14] and this includes acknowledging the histories of colonial extraction. How do we think with anger and with grief?

Where there is anger, then, there must be repentance and reparation so that there can be healing. These reparations can only occur in intersectional ways that understand the conjoined nature of the climate struggle to anti-racist and anti-imperial projects. It is important to note that such conjoined and intersectional struggles already exist. If we look to the movements and actions that have been taking place around the world over the past several decades, we can see that Indigenous, formerly enslaved and anti-colonial movements have been resisting and building alternatives to limitless growth for hundreds of years. This is precisely because the climate crisis, as I have mentioned several times already in this introduction, has long been a reality for many people and ecosystems in the Global South.

A LIBERATION FOR THE EARTH

In the present, it is the destruction caused by cyclones in Mozambique and Zimbabwe, continuing forest fires in the Amazon and Angola. But the centuries-long war economies and forced resource extraction for profit that has led to our current climate crisis were enabled by enslavement and colonization.[15]

As such, anti-colonial resistance has long been interwoven with environmental protection, and we can see that there is a generative rage that binds these struggles together. Indigenous, Black, Global South and racialized communities have long fought with ancestral and innovative knowledge and practice for land rights against large-scale deforestation and resource overexploitation. Often, these struggles have quite literally put bodies on the line, such as in the case of Berta Cáceres. Cáceres was a Lenca Indigenous woman and human rights defender who worked on the front lines defending the territory and the rights of the Indigenous Lenca people. Her organization successfully led a campaign for the defence of the Gualcarque river, which is the site of a proposed dam. Cáceres was assassinated in 2016 by men hired by DESA, the company that was attempting to build the dam.

It is important that theological thinking and faith-based people continue to agitate on these matters, especially to call attention to how coloniality[16] is entrenched within sociocultural, political, economic, gendered and sexual systems of hierarchy, domination, power and exclusion. In this present time, we are also witnessing the emergence of possibilities for epistemological and ontological shift and renewal that can disrupt and push back against nationalist geopolitics, homicidal territorial boundary-policing and divisive categorizations of patriarchal, colonial, neo-colonial and nativist politics. Alongside the politics of exclusion and resentment, we also glimpse possibilities and stagings of resistance, alliances, solidarity and conviviality, politics that are at the heart of documents such as *Laudato Si'* that call for the building of a collective conscience. I was recently struck by Professor Anthony Reddie's comment that not all theologies are liberative. Certainly, this is true, but only if they do not think internationally, only if they do not think for, of and with the global collective, to remember the 'multiplicity of belonging' that arises out of 'an awareness and living of a larger humanity'.[17]

It must be said that faith-based communities and movements do not always pick up these issues of conjoined struggles for justice in their thinking and mobilizing. Akin to their secular counterparts, some faith-based ecological justice movements and writings, navel gazing in grief and unmotivated by rage, can and do sacrifice a hermeneutics of radical justice for the sake of amplifying progressive eschatologies. In doing so, they lend themselves to racialized, imperialistic narratives, often transgressing, transforming and transmuting the rhetoric of discourse on the political left, specifically transforming and silencing narratives of historic suffering. Within these movements, there is a focus on the world that is to come, rather than the world that is, and this focus automatically rejects a hermeneutic of justice. This lack of a justice narrative reinforces and recreates the racialized discourses that already structure the Christian social imagination.[18] The accomplishments of these recent climate movements have relied on an amalgamation of liberal universalism with a kind of anti-politics in sync with contemporary populism. However, their calls to single-issue politics and simplicity of message can easily become a cover for universalizing parochialism. This can exclude, misrepresent, subjugate and silence green movements in the Global South, as well as the politics of land and ecological rights, while professing to act on their behalf.

To focus on some empirics, a number of activists and writers have drawn attention to the lack of diversity in recent environmentalist movements, suggesting that focus has been removed from the ways climate change disproportionately affects communities of colour. This has been exacerbated by groups like XR adopting tactics that exclude migrants and racialized or working-class people by putting them in differentially precarious relationships to state violence. For example, advocating arrest and enjoyment of imprisonment suggests a bewildering short-sightedness regarding the fact that arrest for people from the Global South could mean deportation. And adopting a strategy that considers the police and criminal justice system to be benign, members of which should be worked with as equally oppressed by climate change, masks how they inflict systematic violence and trauma. Additionally, as I note in later chapters, when studying the reception of *Laudato Si'* in North America, the focus within faith-based groups and communities on the role of individual, rather than collective, responsibility, further removes ecological justice from an awareness of a larger humanity. More profoundly, the climate crisis is framed by these forms of environmentalism as a moral problem that affects all of us in equal measure. Shifting from providence to eschatology, their focusing on apocalyptic and extinction narratives is extraordinarily powerful in silencing social differences and conflicts by 'distilling a common threat or challenge to both Nature and Humanity'. The 'people' as political collective are bound together not by shared political views or lived realities, but as victims of a universal crisis.

Without a justice narrative at the heart of environmentalism, it simply becomes vehicles for the privileged – something that development practitioners, now allying themselves to the decolonial turn, are beginning to acknowledge. They are noting that the Sustainable Development Goals are not progressing as imagined because, even with a wide consultation process, recommendations are still being prescribed 'internationally' with a top-down approach. This kind of 'simple' environmentalism is complicit in neo-colonialism and continued oppression in the imperialist core – analogous to asking people to switch off their lightbulbs without considering where the resources to make them were extracted from.

Such thinking preserves a narrative in which the Global North exports technoscientific fixes to the developmental desires of the Global South. Preserving the uni-directional force of development also leans on narratives of protection of supposedly national resources from a multitude outside, to be

INTRODUCTION

safeguarded by an increasingly green military. The formation of a universal 'people' is codified by carbon footprints, net-zero targets and offsetting. These preserve the rationalities, and serve the interests, of the minority Global North. Net-zero emissions targets won't achieve the temperatures required to maintain life in many regions, and green technologies require the acceleration of mineral and metal extraction. They leave uncounted off-shore emissions, imported goods, shipping and aviation, and they require offsetting with tree plantations that reduce biodiversity and require land-grabs, displacing people from homes and livelihoods. So rage becomes important. Not only recognizing the rage that fuels struggles, but also how rage can generate and sustain the justice narrative. There is an urgent need to foreground ongoing battles by communities in the Global South and Indigenous groups in the Global North, to make way within theology and other disciplines for a decolonial, anti-racist vision of ecological struggle. Transforming environmentalism also requires us to consider the ways that knowledge produced within the politics of the Global South challenges the solutions being prescribed in the Global North, to start from 'two bleeding wounds'.[19] For example, many Indigenous writers and scholars have argued that we need to reconsider the relation of politics to land. There exists knowledge that arises from collective memory – from innumerable people living for generations in relation to a specific land and seeing it as the source of all its relations. This is the spiritual, intellectual and emotional dimension of land – the knowledge of co-existence with rivers, streams, air, wind. This further suggests that a spiritual awakening is necessary to address the global ecological crisis.

This is a circuitous way of saying why we need to foreground rage. We do so in order to think differently, collectively, in terms of justice. To think in terms of justice is also to refuse – to refuse in the way the thinker A. M. Kamgieser tells us to, to understand refusal as return – a (re)turning to an (un)known because nothing ever returned to its self-same or absolute – unsettles narratives of resistance that are framed only in opposition. The idea of pouring back or watering expands the conceptualisation of refusal as an act of liberation.[20]

Let us refuse these waves of destruction and pour back in love, in grief, and also rage. Kamgieser and other Indigenous scholars speak beautifully of the desire for restitution, and theologically we cannot approach restitution without reconciliation, without repentance.

The structure of a raging book

The first chapter of this book focuses on the encyclical *Laudato Si'*. The encyclical is a key focal point for my own work as a Catholic theologian, but its importance for a book like this one is its global, collective orientation, a document that is in dialogue with a shared humanity. Drawing from *Laudato Si'*, as well as *Querida Amazonia*, the chapter discusses the concepts of ecological sin and ecological conversion as outlined in these documents. The chapter calls attention to the nexus between Ecology and Liberation, and how, in these encyclicals, the fate of the poor and the planet are repeatedly linked. We see this centring also in the writing of other faith leaders, from the Filipino bishops, to Patriarch Bartholomew, to the Islamic Declaration on Climate Change. By making this examination, the chapter reflects on what is meant by delineating the Earth as the new poor, and how this raises up a call to ecological conversion and to understanding ecological sinfulness in a manner that calls us to rage.

The focus of the second chapter is liberation theology as the guiding framework for the through-argument of this book. The chapter further discusses this concept of the Earth as the new poor, asking how we make continuous our conversion to the preferential option for the poor. The chapter reflects

INTRODUCTION

on the extant work of liberational and decolonial theological voices as they approach a critical understanding of poverty, in particular as we now discuss the Earth as part of that poor, as well as the intersecting crises of all who are at the margins. What does liberation theology tell us about how we might recognize our ecological sin and take up the journey of ecological conversion? How does liberation theology point us to being ecologically converted and, in doing so, remind us of the connection between injustices? What is the dialogue between Leonardo Boff and James Cone?

Chapter 3 focuses the liberation theology lens on contextual, thinking, and, drawing from work in sociology, global studies and critical geography, discusses the history and present of the extraction and colonization that affect and foreground the climate crisis today. What is entangled, extracted, erased? How do these histories compel us to conversion and to confront the ongoing reproduction of colonial, raced and gendered relations from a theological position? Are our theologies adequate for this confrontation? This chapter also draws upon work in critical race, as well as contextual theological work that connects the various peripheries of ecology and race.

In Chapter 4, the historical and theoretical chapters are brought together in an empirical discussion of faith-based ecological justice movements in Canada, Sri Lanka and the United Kingdom, where I have worked and researched. I also give a central place within this to conversations with colleagues working in Christian grassroots movements, some from the Global South, some from Indigenous communities in the Global North, discussing their responses to ideas such as 'ecological sin', 'the Earth as the new poor' and ecological conversion. In these conversations, we also discuss the connects and disconnects between grassroots Christian organizing, the parishes and the academy. I also foreground conversations, papers and discussions of the need for a justice narrative as articulated by Christian activists involved in the struggle for fishing and land rights in Sri Lanka and other parts of South Asia. In these chapters, I attempt to use the empirical studies to demonstrate how anti-racist and anti-imperial struggles are intertwined, and what is missing in those struggles when they are not – that is, when the individual is preferred to the collective.

In the final chapter, I attempt to weave together the previous chapters, their intentions and their tribulations. This chapter reiterates the argument that a lack of a justice narrative has recreated racialized discourses, discourses that are already the 'norm' within our Christian imaginations. In taking up this absence of justice, the chapter then opens the case for a political theology of rage, journeying from Hannah Malcolm's articulations of lament and grief and Boff's discussion of responding to the call of the Earth, and following this grief into what rage, thought of as generative, can do, and why rage is the theology expressed by those most affected by the climate crisis.

Notes

1 https://350.org/rise-from-one-island-to-another

2 S. Miththapala, 2008, *Mangroves*. Coastal Ecosystems Series, Volume 2; Colombo, Sri Lanka: Ecosystems and Livelihoods Group Asia, IUCN, pp. 1–28.

3 J. Watts, 2019, 'Murder, rape and claims of contamination at a Tanzanian goldmine', *The Guardian*, 18 June, www.theguardian.com/environment/2019/jun/18/murder-rape-claims-of-contamination-tanzanian-goldmine, accessed 9.06.2022.

4 Paula Butler, 2015, *Colonial Extractions: Race and Canadian Mining in Contemporary Africa*, Toronto: University of Toronto Press, www.jstor.org/stable/10.3138/j.ctt155jp65, accessed 22.06.2022.

5 Keri Day, 2016, *Religious Resistance to Neoliberalism: Womanist and Black Feminist Perspectives,* London: Palgrave Macmillan.

6 Christina Sharpe, 2016, *In the Wake: On Blackness and Being* Durham, NC: Duke University Press.

7 'An Open Letter to Extinction Rebellion', *Journal of Global Faultlines* 6, no. 1 (2019), pp. 109–12; https://doi.org/10.13169/jglobfaul.6.1.0109, accessed 9.06.2022.

8 Mariana Reyes-Carranza, 'Racial Geographies of the Anthropocene: Memory and Erasure in Rio de Janeiro'; *Politics* (July 2021), https://doi.org/10.1177/02633957211026470, accessed 9.06.2022.

9 Keston Perry, 2021, 'Realising Climate Reparations: Towards

INTRODUCTION

a Global Climate Stabilization Fund and Resilience Fund Programme for Loss and Damage in Marginalised and Former Colonised Societies'; http://dx.doi.org/10.2139/ssrn.3561121, accessed 9.06.2022.

10 Lisa Tilley, '"A Strange Industrial Order": Indonesia's Racialized Plantation Ecologies and Anticolonial Estate Worker Rebellions'; *History of the Present* 10, no. 1 (1 April 2020), pp. 67–83, https://doi.org/10.1215/21599785-8221425.

11 for the wild, 2021, 'Olúfémi O. Táíwò on Climate Colonialism and Reparations / 216' (podcast), *for the wild*, 6 January; https://forthewild.world/listen/olufemi-o-taiwo-on-climate-colonialism-and-reparations-216, accessed 9.06.2022.

12 Leon Sealey-Huggins, '"1.5°C to stay alive": climate change, imperialism and justice for the Caribbean'; *Third World Quarterly* 38, no. 11 (2017), pp. 2444–63.

13 Yolanda Ariadne Collins, 'Racing Climate Change in Guyana and Suriname'; *Politics* (September 2021), https://doi.org/10.1177/02633957211042478, accessed 9.06.2022.

14 Donal Dorr, 2012, *Option for the Poor and for the Earth: Catholic Social Teaching. 20th Anniversary edition*, Maryknoll, NY: Orbis Books.

15 L. Malik, 2019, 'We need an anti-colonial, intersectional climate justice movement'; *AWID*, 3 October, www.awid.org/news-and-analysis/we-need-anti-colonial-intersectional-feminist-climate-justice-movement, accessed 9.06.2022.

16 Coloniality: the 'colonial matrix of power' or the coloniality of power. It is the basis and justification for the exploitation of the world and its resources by European systems of domination. It acts as the foundation for colonialism/imperialism, capitalism, nationalism and modernity. Coloniality naturalizes all of these concepts, or makes them appear universal and inevitable. It can therefore be seen as the governing intellectual ideology of the modern world-system from the sixteenth century to the present day.

17 Mario I. Aguilar, 2021, *After Pestilence: An Interreligious Theology of the Poor*, London: SCM Press, p. 188.

18 J. Kameron Carter, 2008, *Race: A Theological Account*, ACLS Humanities E-Book, Oxford: Oxford University Press.

19 Leonardo Boff and Virgilio P. Elizondo, 1995, *Ecology and Poverty: Cry of the Earth, Cry of the Poor*, London: SCM Press.

20 A. M. Kamgieser, 'Transmediale for Refusal', n.d. Transmediale.de; https://transmediale.de/almanac/to-undo-nature-on-refusal-as-return.

1

The Cry of the Earth: Understanding Ecological Sin and Ecological Conversion

A dying world is ferociously upon us. In 2021, the Intergovernmental Panel on Climate Change (IPCC) laid these facts out for us in stark, cool words. In two decades, the world will reach or exceed 1.5 degrees of warming, and limiting warming even to this requires rapid, transformational change.[1] The extreme weather events we are seeing are unprecedented for any other time in global history. We did not need the IPCC report to tell us this. Depending on where you live in the world we are, almost monthly, shown images of, or are experiencing, increasing wildfires, floods, hurricanes, droughts, cyclones and deadly landslides. In February 2021, in Uttarkhand in India, huge ice blocks that broke off a Himalayan glacier due to rising temperatures triggered a massive flood that killed 83 people and left 121 missing.[2] It was not the first such disaster in the area, nor the last, the deadliest being the flash floods of 2013. Both incidents are linked not only to rising temperatures but also to the construction of hydroelectric projects that destroy not only the local ecology but also affect the well-being of the local community.[3] In perhaps something like divine retribution, the 2021 flood also swept away the Rishi Ganga hydroelectric project. In Latin America and the Caribbean, between 2000 and 2016, nearly 55 million hectares of forest were lost, constituting more than 91% of forest losses worldwide. In the year 2020, a catastrophic fire season over Pantanal burned an area in excess of 26% of the region.⁴ It is a critical time in our history on this Earth, a time for radical reimagining, and is why this book attempts to sit within the spaces of political theology and liberation theology, rather than eco-theology.

That we face a critical environmental moment now must be put into the context of other warnings, other moments of outcry. In 1988, the Filipino Catholic Bishops Conference issued the pastoral letter, 'What is Happening to our Beautiful Land'.⁵ In this document, the bishops focused their concerns for the Earth by crying out against what they saw as an assault on Creation that is 'sinful' and 'contrary' to the teachings of the Christian faith.⁶ Seeing this as a 'critical time' in history, the bishops said:

> To put it simply; our country is in peril. All the living systems on land and in the seas around us are being ruthlessly exploited. The damage to date is extensive and, sad to say, it is often irreversible.⁷

The essence of the document was the bishops' concern that huge plantations and unnatural agricultural practices, as well as deforestation in the name of development, had displaced the natural 'community of the living' that the Creator had ordained. In a very direct fashion, the Bishops Conference reminded their readers that indigenous hunting and food-gathering techniques had shown true connection to nature, to its rhythms and sensitivities, but that much of this had been lost in the name of progress. The pastoral letter is quite striking in the way it is written, because the bishops drew on modern science to argue for a holistic vision of life on Earth. This is intricately tied to the process of Creation, in the discussion of the rainforest for example, showing the true 'integrity of creation'. The appeal is also Christological, calling attention to how Jesus lived on Earth, and that the defacement of Creation also mars the face of Christ. In this the document mirrors the message of *Redemptor Hominis*, in which Pope John Paul II wrote of how the created world recovers its original link with 'the divine source of Wisdom and Love'.[8] In doing so, John Paul II was, according to Dorr,[9] providing the basis for what Deane-Drummond later calls 'deep incarnation'.[10] So, too, the Filipino bishops made this link between the Redeemer and Creation:

> But our meditation on Jesus goes beyond this. Our faith tells us that Christ is the centre point of human history and creation. All the rich unfolding of the universe and emergences and flowering of life on Earth are centred on Him (Eph 1:9–10; Col 1:16–17). The destruction of any part of creation, especially the extinction of species, defaces the image of Christ which is etched in creation.[11]

The letter is also political, it not only laments the destruction of the natural world, it also calls for action on what they see as the ultimate 'pro-life' issue, highlighting not only the new life provided in Christ, but also how Mary calls us to abandon the path towards death, her pain when she sees 'soil erosion, blast-fishing, or poisonous land', and how she calls us to return to the way of Life. The document, in fact, calls for ecological conversion and, through this conversion, for action on a massive scale by individuals, the government and the Church itself.[12] In detailing this, the letter emphasizes again and again the importance of following in the footsteps of tribal ancestors, and also of ongoing efforts such as the Chipko movement in India or the Greenbelt movement in Kenya. They note how local mass resistance is important, citing the Chico project[13] and the Bataan nuclear plant.[14] The letter ends by speaking of what must occur at the three levels of the personal, the ecclesial and the political. At the first level the focus must be on awareness, and also speaking out and organizing on ecological issues. At the ecclesial level it is on calling the community to conversion and developing a Filipino theology of Creation that underpins everything that the Church does. Finally, the letter asks government and non-governmental organizations to resist short-term economic gains at the expense of long-term ecological damage. But it is in the conclusion of this document that the crux of its message lies. In the final few paragraphs, the pastoral letter underscores that the cry of the land and the cry of the people are raised together against 'an exploitative mentality which is at variance with the Gospel of Jesus', but that what is occurring is a situation that the people themselves have brought upon us. Therefore, the necessary response is not only political action, conversion and education, but also realizing our ecological sinfulness. Although the Filipino bishops do not say this explicitly, their constant calling attention to humanity's actions and the degradation of the natural environment and that this goes against God can only be interpreted in this way. There is so much at the intellectual and emotional heart of this document that mirrors the message of *Laudato Si'*, written and promulgated some 27 years later. What is interesting for this book is the understanding within *Laudato Si'* of the fuller development of the idea of the Earth as one with the poor, arguably the 'new' poor, and the more forthright direction on ecological sin, which this writer sees as eminently theo-political categories.[15]

An encyclical from the end of the world

When the encyclical *Laudato Si'* was promulgated, I was working on a Masters of Divinity at a small theological college in Western Canada. For months there had been a frisson of excitement among those of us who 'leaned left' in the Catholic community, knowing that, finally, an encyclical on the environment was to be written. I was also working with the Canadian Catholic Organization for Development and Peace, or Caritas Canada. For that organization, an early attempt to talk about climate justice issues had not fared well, because there had been no apparent papal weight behind the campaign. Now, with this pope from the 'end of the world', there was the promise of an ecological encyclical. The document did not disappoint. It was urgent, energetic, prophetic and passionate.

It demanded a moral imperative. It was written within the tradition and extended the social teaching of the Church. It had profound, emotional passages, and in what is certainly the encyclical's most important passage, it detailed the ecological significance of the Eucharist.[16] The encyclical was anchored in the idea of an integral ecology. What Francis proposes here is consideration of an environmental, economic, social and cultural ecology, as well as what he terms an ecology of daily life.[17] What this means is that we must recognize that there not only exists a link between society and the environment but also a link between our current modes of living, that is, limitless production and consumption, and the 'sinfulness' of our quotidian lives. We were called to an ecological conversion, a spiritual awakening. As a student and also as a part of my own work with Caritas, I came to spend many hours with this particular document, writing on it, teaching with it and reflecting for many long hours on ecological sin. This chapter reflects on ecological conversion and ecological sin as concepts that foreground the Earth as one with the poor. These particular concepts are also key in structuring a political theology of rage.

'The effects of the encounter': on ecological conversion

Many column inches, podcasts, books and journal articles have been devoted to understanding what 'ecological conversion' is. Fundamentally, it is a pastoral issue. As Celia Deane-Drummond has commented, Pope Francis is, at heart, a priest,[18] and this innate pastorality is reflected in the encyclical *Laudato Si'*. Indeed, there is something in reading the encyclical that makes one feel called to an examination of conscience. Listen to the cry of the Earth and of the poor. Reject the throwaway culture of consumerism. Commit to the perusal of an integral ecology. An integral ecology flows from an ecological conversion. In calling us to conversion, Pope Francis is in fact underscoring the deeply traditional nature of *Laudato Si'* within the Catholic tradition, especially within Catholic Social Teaching (CST). The heart of CST is in fact this conversion to justice which is a necessary transformation, either personal or societal, which then reorients our existing concerns so that we work towards changing sinful socio-economic structures. Sinful structures are defined by John Paul II in *Solicitudo Rei Socialis* as those that are 'rooted in personal sin and thus linked to concrete acts of individuals who introduce [them], consolidate them and make them difficult to remove' (SRS 36).[19] Francis focuses on our disposable culture as sinful and calls our attention to the need to overturn it. These are calls we encounter in Paul VI's *Populorum Progressio* in its description of integral human development, in particular the urging of a process of change to considering the whole human person and the need for an openness to love, friendship, prayer, contemplation. Francis revisits *Populorum Progressio* when he builds his understanding of ecological conversion by picking up on the need to grow in love by looking at the mystical meanings in leaves, dewdrops and mountains.[20] The particular phrasing of 'ecological conversion' comes from later papal teaching, that is, from John Paul II, a pope who was certainly more at the conservative end of the spectrum compared to his Argentinian successor.

Indeed, as many have noted, the phrase 'ecological conversion' was first referred to in a 2001 papal audience when John Paul II was providing a theological analysis of our duty towards Creation. John Paul II, referencing his encyclical *Evangelium Vitae*, reminded his listeners that humanity's stewardship of the Earth must reflect the lordship of the Divine, the quality of life and ecology. Very much like the Filipino bishops, John Paul II saw this not only as an orientation to ecology but also as part of the believer's commitment to Christ – Christ who provides the pattern for right relationships through the Divine's affirmation of the natural world that is enfleshed in the Incarnation. Deane-Drummond[21] has noted that John Paul II was very direct in this use and teaching on ecological conversion, with particular consideration to his instruction to bishops that they must teach their flock about the correct relationship of human beings with nature. Deane-Drummond sees this conversion as linked both to Christology and to the doctrine of God.[22] As such, it is from John Paul II that we have this understanding that ecological conversion is part of the conversion to Christ, to a cosmic Christology:

> We must therefore encourage and support the 'ecological conversion' which in recent decades has made humanity more sensitive to the catastrophe to which it has been heading. Man is no longer the Creator's 'steward', but an autonomous despot, who is finally beginning to understand that he must stop at the edge of the abyss.[23]

The concept of ecological conversion also grounds the climate justice writings and thoughts of Patriarch Bartholomew, someone with whom Francis is in dialogue. The Patriarch notes that the fundamental reorientation that we must undergo is a spiritual one, one that turns us to deeper repentance and a willingness to forgo.[24] The encounter is also, for the Patriarch, rooted in the experience of being believers:

> When we are transformed by divine grace, then we discern the injustice in which we are participants; but then we will also labour to share the resources of our planet; then, we realize that eco-justice is paramount – not simply for a better life, but for our very survival.[25]

While we do, then, find the phrasing of ecological conversion and the framework surrounding it in previous documents and teachings, it is in *Laudato Si'* that the concept receives its full treatment. Francis grounds this conversion in the faith that we proclaim: that we have been saved by Jesus Christ. This profound encounter must be made evident in our actions, and it is to this we are summoned. As the document is a Catholic one, this encounter is very deeply focused on the eucharistic meal. Inasmuch as Christ's body is blessed, broken and shared, so too are we called to become that which we receive.[26]

> The Eucharist joins heaven and earth; it embraces and penetrates all creation. The world which came forth from God's hands returns to him in blessed and undivided adoration: in the bread of the Eucharist, 'creation is projected towards divinization, towards the holy wedding feast, towards unification with the Creator himself' [Benedict XVI, Homily for the Mass of Corpus Domini (15 June 2006): AAS 98 (2006), 513]. Thus, the Eucharist is also a source of light and motivation for our concerns for the environment, directing us to be stewards of all creation.[27]

Becoming what we receive entails working for justice.[28] The encounter anoints us, it sends us out into the world with a gaze that is renewed. This is the gaze of Francis of Assisi, one that is alive with awe and wonder. We contemplate the beauty of Creation, thus moving our gaze from the technocratic to one that is joyful and contemplative. This new 'gaze' is Christ-like[29] and in this way we are charged to make a paradigm shift in which we, individually and communally, must change at all levels and become conscious of a new awareness of seeking to be in right relationship with God, neighbour and Creation. To be thus awe-struck is to have our eyes open to the mystery of being and of the Divine.

> If we approach nature and the environment without this openness to awe and wonder, if we no longer speak the language of fraternity and beauty in our relationship with the world, our attitude will be that of masters, consumers, ruthless exploiters ... By contrast, if we feel intimately united with all that exists, then sobriety and care will well up spontaneously.[30]

This language in *Laudato Si'* draws on the very spiritual idea of conversion. This is the reorientation, the new gaze or perspective that then informs our goals as well as our choices and action. It is a turning point, but also a process of turning away and towards.[31] It is a radical break away from the status quo, as well as the moralities that we are used to. Ileana Porras agrees,[32] arguing that the importance of this kind of religious conversion is that we are challenged as well as empowered to undergo a process of self-sacrifice that asks us to make this radical break from our ways of living, and the ways in which we produce and consume, towards the work of demanding justice now. These are oft-repeated motifs not only in *Laudato Si'* but throughout the theology of Francis' pontificate. In *Laudato Si'*, as well as in other documents within his papal teaching, Francis identifies two dominant social structures which he finds to be the enablers of the 'sinful' throwaway culture of our times. The first is the unfettered free-market economy and the second is the technocratic paradigm. For Francis, the continuation of extreme poverty and growing inequality are scandalous and sinful. The effects of the encounter in Christ must mean that we have a personal, political and ecclesiological imperative to overturn this.[33] Others[34] have written significantly on how Francis' theology understands these two structures. What I will focus on here is what Francis says is necessary in order to engender decisive action. This, of course, is ecological conversion, whereby we take on a profound interior, as well as community conversion so that we address the collective challenges of ecological injustice. This is what is important for the 'radical break'. It is an interruption of the social structures within which we live and a shift to imagining otherwise. This, essentially, is what is meant by integral ecology, the realization of the interrelation of everything. This is beyond simply an idea of sustainable development, as Porras argues,[35] but is suggesting a new sensibility that shifts away from the purely anthropocentric. Porras explicitly sees it as anti-anthropocentric,[36] but I would prefer to see it as the more holistic and important recognition that it is for our conversion towards the understanding of multiple and interlocking networks of relationships. Therefore it is a move away from an exclusively anthropocentric view that underscores the argument that Nature must not be seen only as of utility for humankind. This develops fully within Francis' theology, which not only demands the pursuit of integral ecology, but also demands justice between the generations. We must hear both the cry of the Earth and the cry of the poor. Ecological conversion is thus the first step towards working for an integral ecology, and towards integral human development as well. *Laudato Si'* as well as the more recent *Querida Amazonia* see a deep connection within all of Creation – the social, economic, cultural, political and ecological dimensions; in order to address environmental degradation, one must also address human and social degradation. The moment of ecological conversion is also a social approach that demands engagement with justice. For Deneulin, this is why *Laudato Si'* provides an important teaching for international development,[37] particularly as a guide for questioning power and privilege, and is also an extension of liberation theology and Catholic Social Teaching, linking personal conversion and the struggle for structural change. Ecological conversion must happen at all levels of society. Repeatedly, we must be aware that what is asked of us is a profound transformation of the relationship that, individually and communally, we have to others and to the Earth, and the rejection of narratives of domination and exploitation. The process is not immediate, but the 'effect of the encounter' must be that constant effort is made in a long journey of transformation.

In noting an aspect of journeying in ecological conversion, Zhang draws the line between Francis of Assisi and Pope Francis, noting that this kind of spiritual conversion linked to reparation and restoration is in the spirit of the former's idea of the conversion experience.[38] Once again, it is the moment of encounter that charges this conversion. Zhang narrates, for example, Francis of Assisi's encounter with a leper, his encounters with the natural world, and then his openness to the beauty of the Divine as he sees it all around him. As Omerod and Vanin note, this very closely mirrors Lonergan's idea of conversion, in that the effect of the encounter is not only to love the Divine, but also to love all that the Divine loves, and to love as the Divine loves.[39] Hence, ecological conversion suggests a state of being that pushes our love into a cosmic dance, into a '[sublation] with an all encompassing good'.[40] Ecological conversion recognizes the sacramentality of a world in which we understand that all of Creation is revelatory; it suggests to us aspects of the Divine. In this way ecological conversion is a reminder of pan-en-theism, that God is in all things. Sacramentally through the Eucharist as well as through experience of the encounter, we are in connection to the living world. We recognize its sacredness and long for a greater intimacy with it, for a covenant relationship. Again and again, the thundering effect of the encounter, of the reorientation to Christ, is entering into this passionate love for Creation. As such, it is a conversion that is moral, intellectual and religious.

Ecological conversion is also Francis' imperative to work towards the development of a spirituality of global solidarity. Indeed, the rather remarkable thing about the encyclical is how decentralized it is, how much it explicitly brings together a crowd of voices. Francis quotes from the ecological work of Patriarch Bartholomew, as mentioned, but also from the writings of bishops from Mexico, Bolivia, Paraguay, Argentina, North America, New Zealand, Japan, Portugal, South Africa and the Dominican Republic. Underscoring what has been noted above regarding the interconnectedness of relationships, in doing this we see ecological conversion being developed theologically through a concern for real-life justice and problems 'on the ground'.[41] As Campos argues,[42] this is a recognition of theology that is born out of human experience, especially that of suffering. Thinking with the concept of *dukha*,[43] Campos notes the importance to *Laudato Si'* of knowing the suffering of the other. Knowing the reality of such suffering is also part of the process of ecological conversion, and, for Campos, this groundedness is Francis asking us to be provoked towards a prophetic rage that stands against those who would perpetrate violence and injustice, a theology that resists evil and suffering and is imbued with generative rage. In fact, in *Querida Amazonia*, we see Francis directly rejecting an ecological justice movement that does not take into account the experiences and struggles of those who have historically been at the receiving end of colonialism, and who now fight against the rapaciousness of multinational corporations. Indeed, at this point in the exhortation, and in what he sees as a first step towards building solidarity and dialogue, Francis calls for righteous anger:

> We need to feel outrage, as Moses did (cf. Ex 11:8), as Jesus did (cf. Mk 3:5), as God does in the face of injustice (cf. Am 2:4–8; 5:7–12; Ps 106:40). It is not good for us to become inured to evil; it is not good when our social consciousness is dulled before 'an exploitation that is leaving destruction and even death throughout our region ... jeopardizing the lives of millions of people and especially the habitat of peasants and indigenous peoples'.[44]

Directly following this passage in *Querida Amazonia*, Francis notes a need for the Church to ask forgiveness, and in doing so reminds us of the importance of acknowledging sins that have been committed.

> At the same time, since we cannot deny that the wheat was mixed with the tares, and that the missionaries did not always take the side of the oppressed, I express my shame and once more 'I humbly ask forgiveness, not only for the offenses of the Church herself, but for the crimes committed against the native peoples during the so-called conquest of America' as well as for the terrible crimes that followed throughout the history of the Amazon region.[45]

This explicit focus on sin is also tied to the process of ecological conversion. That is to say, to go through the process of ecological conversion is also to acknowledge ecological sin.

The sins against our common home

If you google the phrase 'ecological sin' the results will mostly come from a variety of news sources that tell you that the Catholic Church is considering introducing sins against the ecology as part of the formal Catechism. Should ecological sin be introduced into the examination of conscience, it would be more than simply recognizing the misuse of the Earth, but would include our repeated failure to struggle for justice and recognize the suffering of the Other. When Francis speaks of sinfulness in the encyclical, he references much of the understanding of integral ecology that I have mentioned above. Sin is the breakdown of our relationships with God, with the Earth, with each other and with the created order. Irwin notes that sin itself is only referenced four times in the encyclical, once in relation to sin in Genesis, then to discuss the visions of Francis of Assisi and St Bonaventure and then to speak about throwaway culture.[46] Francis does not explicitly use the phrase 'ecological sin' in the encyclical, but it is clear throughout that the document is calling us to think deeply about sinfulness, sinfulness caused by turning away from the 'right relationship' with Creation to which Christ himself witnessed.

> The creation accounts in the book of Genesis contain, in their own symbolic and narrative language, profound teachings about human existence and its historical reality. They suggest that human life is grounded in three fundamental and closely intertwined relationships: with God, with our neighbour and with the earth itself. According to the Bible, these three vital relationships have been broken, both outwardly and within us. This rupture is sin.[47]

> Creation us allowed/in intimate love/to speak/to the Creator/as if to a lover.
> Creation/is allowed/to ask/for a pasture/homeland.
> Out of the Creator's fullness/this request is granted to Creation.[48]

The importance of relationality is paramount to both conversion and sin, because the way back to right relationship, to living an integral ecology, is the process of ecological conversion. In discussing the importance of these relationships, Francis continues to emphasize the collective over the individual. We are not supposed to understand our sinfulness as only limited to the personal sphere, but also to think about societal structures of sin, what Cardinal Turkson terms an 'understanding of sin with a planetary perspective'.[49] Ecological sinfulness breaks the threefold relationship between the individual or community and God, neighbour and the Earth, both outwardly and within themselves, as the above quotation from paragraph 66 notes. If sin is a kind of self-assertive freedom, then the quest for power and taking the place of God in trying to dominate and subjugate the natural world is wrong. The throwaway culture that must be rejected in our conversion experience is one that denies the experience and suffering of others, and engagement in this culture, as well as the denial of others' *dukha*, is sinful. This sinfulness has engendered a spiritual crisis because humanity has self-asserted itself against the Creator. Hence the very logical connection between the holistic understanding of conversion described above and what sinfulness is.

> The violence present in our hearts, wounded by sin, is also reflected in the symptoms of sickness evident in the soil, in the water, in the air and in all forms of life. This is why the earth ... is among the most abandoned and maltreated of our poor.[50]

In developing this idea of ecological sin, Francis gives a central place to the thinking of Patriarch Bartholomew, who points to several sins against Creation. Bartholomew lists these as the destruction of biological diversity, the degradation of the integrity of the Earth, the stripping of the Earth of natural forests and wetlands, and the contamination of water, land, air and life.[51] Francis himself, while Archbishop of Buenos Aires, had commented in the fifth Aparecida document on the sinfulness of transnational companies that subordinate local peoples, local economies and local governments, as well as that of the international extractive industries and agribusinesses that damage biodiversity and exhaust natural resources.[52] This is in many ways very reminiscent of the pastoral letter of the Filipino bishops with which this chapter opened. These abuses against the environment, as well as against Indigenous peoples and local ways of being, are themselves sins against the Creator, a turning away from right relationship. For many scholars, this foregrounding of sinfulness in this way highlights the encyclical's debt to liberation theology.[53] As the next chapter deals with liberation theology in more detail, it is enough here to mention that in thinking of what ecological sinfulness *is*, we can see that Francis, following in the long line of local bishops' conferences and both faith-based and secular social justice activists, is naming the dominant paradigms of economic liberalism and neo-liberalism as the fundamentals of this sinful situation. Exponential, unlimited growth that breaks the natural cycles of the world continually sins against the Creator; it cannot pretend to be ethical.[54] Indeed, in the encyclical we see Francis giving a central place to Catholic Social Teaching on private property and reminding us of the sinfulness of seeing the right to private property as absolute or inviolable.[55] Indeed, he goes on to elaborate on the fact that due to the understanding of the common good, ownership is not for the individual, neither is wealth to be exclusive; it can only be managed for all. Private property as an absolute principle is sin. Francis' example here is the privatization of water. As water is integral to human and animal survival, to bring it into privatization due to the principles of liberal economics breaks the relationship decreed by the Creator. We are called by this to repentance, and also to refuse the projected ethicality of the neo-liberal system. To do this work of repentance, of understanding and acknowledging our sinfulness, to engage in ecological conversion and journey towards an integral way of living, we also require that prophetic rage that Francis suggests in *Querida Amazonia*. In my final chapter, I pick this up and discuss the framework for a political theology of rage. However, as with the encyclical, the structure of theological rage echoes down to us through Scripture, through the active cries of the historically excluded, and also in the theologies coming out of the Global South. In the next chapter, I will discuss this by discussing what is elaborated within liberation theology.

A pause for acknowledgement

Before I begin the discussions that follow in the next chapters, it is important that I acknowledge that both *Laudato Si'* and the theology of Pope Francis do contain what academics would term 'gaps'. Certainly, a lacuna exists in the document in that it does not acknowledge the disproportionate harm that women experience, especially when engaged in environmental activism, and when at the receiving end of extreme weather events. Indeed, in June 2021, the Laudato Si' Institute organized a special conference that remarked on this particular gap in the document and aimed to fill it by discussing questions of gender injustice in terms of social and environmental justice. As documents from within the UN and the Caritas networks have noted, women often have far fewer resources with which to cope with both poverty and extreme climate events.[56] Even if the problems of water pollution and water shortage, which the encyclical mentions several times (see LS 2, 8, 20, 24, 27–31, 35, 37, 40, 44, 48, 72, 140, 164, 185, 211, 234, 235), affect all of the population in some parts of the globe, they represent a particular weight upon women's shoulders, since 'in many countries, women are responsible for finding and collecting water for their families. All the water they need for drinking, washing, cooking, cleaning. They walk miles, carry heavy burdens, wait for hours and pay exorbitant prices.'[57] Toldy finds that the non-existence of women in the encyclical impedes its aim of building an 'integral ecology'. As Toldy notes, this is particularly important when we consider how environmental degradation has a direct impact on women's reproductive health.[58] Toldy, and also Tina Beattie, find it problematic that Francis does not critique the androcentrism of environmental degradation. Both argue that the dominating nature that is marked out as sinful in *Laudato Si'* is part of a structural order in which women are seen as beings to be dominated, and upon whom violence is often directed. An 'integral' encyclical can and should also contain a radical critique of androcentrism. Schlichten underscores all these arguments by saying that the problem with the encyclical is that while it contains a deep anti-capitalist critique, it does not go far enough.[59] I would agree that, as already mentioned, the encyclical does miss the opportunity, as aforementioned, to link capitalism to patriarchy. There is a deep sexism to environmental exploitation that requires foregrounding and challenging in any movement towards ecological justice.

Kevin O'Brien further notes the problematic way in which Francis maintains the traditional Roman Catholic understanding of the family, with no attempt to think beyond the binary of two genders.[60] Agnes Brazal also finds the centrality of the family difficult, arguing that the patriarchal emphasis of this association refuses a moral imperative for men to disengage from the hierarchy of an all-powerful father figure.[61] O'Brien also reminds us of the argument from Ivone Gebara,[62] who sees Francis' understanding of women as charitable and compassionate but not empowering. Nichole Flores too is troubled by the ecological vision of the family in *Laudato Si'* because it continues the gender subjugation that is innate within Catholic teaching.[63] Sharon Bong, reading the encyclical from an Asian feminist theological viewpoint, shows appreciation for the discussion within the text of the differentiated responsibilities between the Global North and Global South, but struggles as Todly does with the encyclical's inability to synergize climate justice and gender justice.[64] Bong argues that there is a need for such a document to discuss the differences that matter, such as age, class, caste, ethnicity, gender and religious affiliation, as these differences also impact the struggle with poverty. I agree with Bong here, and also join in her critique of Francis' characterization of the Earth as a fragile female entity that is in constant need of male protection. Bong notes that this particular approach also ignores the theologizing that Asian feminist religious scholars have done by engaging the work of female activists who have worked hard to challenge traditional Asian understandings of the woman as self-sacrificial and docile. Neither women nor the poor are passive victims, an aspect that I dwell on significantly in Chapters 4 and 5, and why I argue for theological rage. Susan Rackozy furthers these critiques by calling for the Catholic Church to add the missing element of ecofeminism to teaching on ecological justice.[65] For Rackozy, ecofeminism's stress on the sacredness of Creation, the non-dualistic interpretation of matter and spirit and the importance of the web of life correlate with Francis' underscoring of the environment as a locus of the Divine presence.

Conclusion

An incomplete document can still be instructive and inspiring. The strengths and the excitement of *Laudato Si'* are found in the remarkable ways in which ecological conversion is developed, and in the ways in which the pope carries on the radical traditions of Catholic Social Teaching to name and resist unfettered growth. It is also an encyclical that calls attention to a variety of global voices and that carries on the unfinished work of popes like John XXIII and Paul VI, and in doing so opens out traditional Catholic teaching in a way that Francis' immediate predecessors were unwilling to do. It is a document that is willing to bring liberation theology into the 'mainstream' of Church teaching. It is a document in dialogue, which offers itself from the beginning as part of a dialogue with all peoples. It is a document with clear pastoral priorities which reflects on and seeks to give a central place to realities on the ground.[66] Or, shall we say, it encourages theologians and postulant voices such as myself to speak from and within 'theologies of the ground'. It is these theologies – those of nuns at the front lines of protest, of trade unionists and fishing rights activists – that I engage with in developing a political theology of rage, guided by liberation theology as the lens through which to do so.

Notes

1 Intergovernmental Panel on Climate Change, 2021, 'Climate Change Widespread, Rapid, and Intensifying', *IPCC*, 9 August, https://www.ipcc.ch/2021/08/09/ar6-wg1-20210809-pr/, accessed 9.06.2022.

2 Victoria Elms and Natasha Muktarsingh, 2021, 'Uttarakhand Dam Disaster: What Caused India's Deadly Flood?', 16 February, *Sky News*, https://news.sky.com/story/uttarakhand-dam-disaster-what-caused-indias-deadly-flood-12214731, accessed 9.06.2022.

3 N. Rana, S. P. Sati, Y. P. Sundriyal et al., 'Socio-economic and environmental implications of the hydroelectric projects in Uttarakhand Himalaya, India', *Journal of Mountain Science* 4 (2007), pp. 344–53, https://doi.org/10.1007/s11629-007-0344-5, accessed 9.06.2022.

4 WMO, 2021, Review of *State of the Climate in Latin America and the Caribbean*, Geneva.

5 I was very struck when writing this by how commentators note that the title of the pastoral letter is often mis-written as ending with a question mark when, in fact, it does not. The lack of interrogation makes the statement more direct, as if to say, 'This, this devastation is what is happening to our beautiful land.' It is not, therefore, a document of lamentation.

6 In zeroing in on this particular pastoral letter, I am not attempting to say it is alone of its type or that it had no precedents. As Sigmund Bergmann has noted, for example, Gregory of Nazianzus was writing about the link between extreme weather and social injustice in the fourth century. In 1971, *Octogesima advenesis* had a note regarding the risks that come from 'ill-considered exploitation of nature'. Also in 1971, the document 'Justice in the World' called on the rich to accept a less material way of life and made a direct connection between ecology and justice. Donal Dorr argues that this might be one of the first instances in which a connection between the option for the poor and the option for the Earth is made. I am grateful both to Dorr's work and to a semester of independent study during my MDiv with Professor Bob McKeon that allowed me to read and understand the long history of the Catholic Church's thinking on ecology, as well as the significant volumes of extant work on eco-theology. I begin this chapter with a focus

on the Filipino bishops' pastoral letter because it is a testament from the Global South and also because of the political nature of its writing.

7 Catholic Bishops Conference of the Philippines (CBCP), 1988, 'What Is Happening to Our Beautiful Land', available from http://www.inee.mu.edu/documents/19WhatisHappeningtoOurBeautifulLand_000.pdf.

8 John Paul II, 1979, *Redemptor hominis*, Vatican: Libreria Editrice Vaticana, §8.

9 Donal Dorr, 2012, *Option for the Poor and for the Earth: Catholic Social Teaching*. 20th Anniversary edn, Maryknoll, NY: Orbis Books.

10 Celia Deane-Drummond, 'Joining in the Dance: Catholic Social Teaching and Ecology', New Blackfriars 93, no. 1044 (2012), pp. 193–212; and 2017, *A Primer in Ecotheology: Theology for a Fragile Earth*, Eugene, OR: Cascade Books.

11 CBCP, 'What Is Happening to Our Beautiful Land', p. 7.

12 Seán McDonagh, 'What Is Happening to Our Beautiful Land?' *The Furrow* 39, no. 7 (1988), pp. 472–4, http://www.jstor.org/stable/27678747, accessed 10.06.2022.

13 This is in reference to mass resistance by the Kalinga and the Bontok that stopped the construction of the Chico dams, which threatened to displace thousands of people. These dams were World Bank-funded priority projects of the dictator Marcos in the late 1970s and early 80s ('The Cordillera Resistance against Chico Dam and Cellophil', *Environmental Justice Atlas*, https://ejatlas.org/conflict/indigenous-peoples-resistance-against-cellophil-resource-corporation).

14 After the 1973 world oil crisis, the Philippines began looking to nuclear power as a source of domestic energy. This led to an anti-nuclear movement throughout the late 1970s and 80s to stop the construction of nuclear facilities and rid the country of US military bases allegedly housing nuclear weapons. The focal point of the movement was to take down the Bataan Nuclear Power Plant (BNPP), 100 km west of Manila (source: World Environment Atlas).

15 The phrase 'poor' is used often in this book and, as such, there is a need to have some definition of what the author understands as 'the poor'. This is poverty understood as multidimensional. That is, the state of being in which an individual or a community or an entity can be facing multiple disadvantages at the same time, concerning their income, gender, geographic location, caste, class, race, creed, psychological difficulties, access to services, access to justice, ability for their knowledge to be acknowledged, their intergenerational trauma, their multiple displacement. This list is not an exhaustive one. The journey 'out' of poverty is not a linear one. If the poor are always with us, it is

because our ways of socio-economic living continually create, maintain and reproduce systems of marginalization, exclusion and silencing.

16 Pope Francis, 2016, *Encyclical letter Laudato Si' of the Holy Father Francis, On Care for Our Common Home*, Vatican: Libreria Editrice Vaticana, §236.

17 LS §147.

18 Deane-Drummond, *A Primer in Ecotheology*, p. 61.

19 John Paul II, 1988, *Encyclical letter Sollicitudo rei socialis of the supreme pontiff, John Paul II, to the Bishops, Priests, Religious Families, sons and daughters of the Church and all people of good will for the twentieth anniversary of Populorum Progressio*, Vatican: Libreria Editrice Vaticana.

20 LS §233.

21 Celia Deane-Drummond, 2018, 'Catholic Social Teaching and Ecology: Promise and Limits', in William T. Cavanaugh, *Fragile World: Ecology and the Church*, Eugene, OR: Wipf and Stock Publishers.

22 Deane-Drummond, 'Catholic Social Teaching and Ecology', p. 64.

23 John Paul II, 2001, 'General Audience', *The Holy See*, 17 January, https://www.vatican.va/content/john-paul-ii/en/audiences/2001/documents/hf_jp-ii_aud_20010117.html, accessed 30.10.2021.

24 John Chryssavgis, 2011, *On Earth as in Heaven: Ecological Vision and Initiatives of Ecumenical Patriarch Bartholomew*, New York: Fordham University Press.

25 Ecumenical Patriarch Bartholomew, 2008, *Encountering the Mystery: Understanding Orthodox Christianity Today*, London: Doubleday Books.

26 Michael Edomobi, 2018, 'Eucharist as Fruit of the Earth and the Environmental Degradation in the Niger Delta: Implication of Louis Mariechauvet's Tripartite Cycle of Gift, Reception and Return Gift for an Ecological Conversion', *Jesuit School of Theology Dissertations* 21, available from https://scholarcommons.scu.edu/jst_dissertations/21, accessed 10.06.2022.

27 LS §236.

28 Andrew Casad, 2013, 'The Eucharist and the Right Relationship to the Land', *Catholic Rural Life*, https://catholicrurallife.org/the-eucharist-and-right-relationship-to-the-land-by-andrew-casad/, accessed 28.06.2022.

29 LS §§97, 99.

30 LS §11.

31 Ileana M. Porras, '*Laudato Si'*, Pope Francis' Call to Ecological Conversion: Responding to the Cry of the Earth and the Poor – Towards an Integral Ecology', *AJIL Unbound* 109 (2015), pp. 136–41.

32 Porras, '*Laudato Si*''.
33 Lisa Sowle Cahill, 'Laudato Si': Reframing Catholic Social Ethics', *Heythrop Journal* 59, no. 6 (2018), pp. 887–900.
34 For example, Dennis O'Hara, Matthew Eaton and Michael T. Ross, 2019, *Integral Ecology for a More Sustainable World: Dialogues with* Laudato Si', Lanham, MD: Lexington Books; Joshtrom Isaac Kureethadam, 2016, *Rebuilding Our Common Home: Ten Green Commandments of* Laudato Si', Bengaluru, India: KJC Publication.
35 Porras, '*Laudato Si*''.
36 Antonio Porras, 2017, 'Virtues for an Integral Ecology', in A. Sison, G. Beabout, I. Ferrero (eds), *Handbook of Virtue Ethics in Business and Management*, Dordrecht: Springer Dordrecht, https://doi.org/10.1007/978-94-007-6510-8_130, accessed 10.06.2022.
37 Séverine Deneulin, 'Religion and development: integral ecology and the Catholic Church Amazon Synod', *Third World Quarterly* 42 no. 10 (2021), pp. 2282–99.
38 Xue Jiao Zhang, 'How St. Francis Influenced Pope Francis' Laudato Si'', *CrossCurrents* 66, no. 1 (March 2016), pp. 42–56, https://www.jstor.org/stable/26605735, accessed 10.06.2022.
39 Neil Ormerod and Cristina Vanin, 'Ecological Conversion: What Does It Mean?', *Theological Studies* 77, no. 2 (June 2016), pp. 328–52, https://doi.org/10.1177/0040563916640694, accessed 10.06.2022.
40 Ormerod and Vanin, 'Ecological Conversion', p. 334.
41 Donal Dorr, 'The Ecology Encyclical – the Care of Our Common Home', *The Furrow* 66, no. 7/8 (July/August 2015), pp. 387–93, http://www.jstor.org/stable/24636154, accessed 10.06.2022.
42 Clement Campos, 'Laudato Si': An Indian Perspective', *Theological Studies* 78, no. 1 (March 2017), pp. 213–25, https://doi.org/10.1177/0040563916682428, accessed 10.06.2022.
43 A concept within Hinduism and Buddhism meaning strife, suffering or the experience of sadness.
44 Pope Francis, 2020, *Post-synodal Apostolic Exhortation Querida Amazonia of the Holy Father Francis to the People of God and to all Persons of Good Will*, Vatican: Libreria Editrice Vaticana, §15.
45 *Querida Amazonia*, §19.
46 Kevin W. Irwin, 2016, *Commentary on Laudato Si': Examining the Background, Contributions, Implementation and Future of Pope Francis's Encyclical*, Mahwah, NJ: Paulist Press.
47 LS §66.
48 Gabriele Uhlein and Hildegard of Bingen, 1984, *Meditations with Hildegard of Bingen*, Santa Fe, NM: Bear & Co., p. 67.
49 Peter Kodwo Appiah Turkson, 2018, '"Integral Ecological Conversion" Cardinal Peter Kodwo Appiah Turkson', *Laudato Si' Institute*,

https://www.laudatosiinstitute.org/integral-ecological-conversion-cardinal-peter-kodwo-appiah-turkson/, accessed 10.06.2022.

50 LS §2.

51 Patriarch Bartholomew, 1997, 'Address at the Environmental Symposium, Saint Barbara Greek Orthodox Church, Santa Barbara, California, November 8, 1997'.

52 J. Aparecida Bergoglio, 2007, 'Documento Conclusivo: V Conferencia General Del Episcopado Latinoamericano Y Del Caribe', Bogotá: CELAM, p. 66.

53 Philip Goodchild, 'Creation, Sin, and Debt: A Response to the Papal Encyclical *Laudato si'*, *Environmental Humanities* 8, no. 2 (2016), pp. 270–76.

54 Goodchild, 'Creation, Sin, and Debt'.

55 LS §93.

56 UN Women, 'In Focus: Women and Poverty', *UN Women. The Beijing Platform for Action Turns 20*, https://beijing20.unwomen.org/en/in-focus/poverty#facts, accessed 10.06.2022.

57 Teresa Toldy, 2016, 'Reading the Encyclical Letter "Laudato Si'" with Gender Lens', *CIDSE*, https://www.cidse.org/2016/07/25/reading-the-encyclical-letter-laudato-si-with-gender-lens/, accessed 10.06.2022.

58 Teresa Toldy, 'Someone is missing in the Common House: The Empty place of women in the encyclical *Laudato Si'*, *Journal of the European Society of Women in Theological Research* 25 (2017), pp. 167–89.

59 David von Schlichten, 'Strange as this Weather has been: Teaching Laudato Si' and Ecofeminism', *Journal of Moral Theology* 6, Special Issue 1 (2017), pp. 159–68.

60 Kevin J. O'Brien, 'The Scales Integral to Ecology: Hierarchies in *Laudato Si'* and Christian Ecological Ethics', *Religions* 10, no. 9 (2019), p. 511, https://doi.org/10.3390/rel10090511, accessed 10.06.2022.

61 Agnes M. Brazal, 'Ethics of Care in Laudato Si': A Postcolonial Ecofeminist Critique', *Feminist Theology* 29, no. 3 (May 2021), pp. 220–33, https://doi.org/10.1177/09667350211000614.

62 Ivone Gebara, 2017, 'Women's Suffering, Climate Injustice, God, and Pope Francis's Theology: Some Insights from Brazil', in Grace Ji-Sun Kim and Hilda P. Koster (eds), *Planetary Solidarity: Global Women's Voices on Christian Doctrine and Climate Justice*, Minneapolis, MN: 1517 Media, pp. 67–80.

63 Nichole M. Flores, 'Our Sister, Mother Earth: Solidarity and Familial Ecology in Laudato Si'', *Journal of Religious Ethics* 46 (2018), pp. 463–78.

64 Sharon A. Bong, 2017, 'Not Only for the Sake of Man: Asian Feminist Theological Responses to *Laudato Si'*', in Grace Ji-Sun Kim

and Hilda P. Koster (eds), *Planetary Solidarity: Global Women's Voices on Christian Doctrine and Climate Justice*, Minneapolis, MN: 1517 Media, pp. 81–96.

65 Susan Rakoczy, 'Is Pope Francis an Ecofeminist?' Transformation: Where Love Meets Social Justice. Online text no longer available.

66 Mario I. Aguilar, 2014, *Pope Francis: His Life and Thought*, 1st edn, Cambridge: The Lutterworth Press, https://doi.org/10.2307/j.ctt1cgdz3p, accessed 10.06.2022.

2

All These Bleeding Wounds – Liberational Theologies and the Ecological Crisis

> I think in order to really heal the world we need the 'wisdom of darkness.' This can be the Third World, dark people, women, or our 'shadows,' ... all the things we do not want to confront within ourselves, so we project them onto others and call them terrorists. So, I think that we need 'endarkenment' for a while, not enlightenment, to heal the world.
> (Chung Hyun Kyung)[1]

In Amos 5.21–25, the God of Moses and of Abraham thunders at us, 'I hate, I despise your religious festivals ... Away with the noise of your songs!' In this passage, easily one of the most often quoted among 'social justice' theologians, this God asks instead for the offerings of justice and righteousness. Indeed, we may read this portion of Scripture as an indication of the relationship between faith and praxis. It is this relationship that is the fundamental tenet of liberation theology.

This is going to be a chapter of many moving parts. In the main, it discusses the arguments from liberation theology, the guide that such theology presents for linking faith and praxis. As such, this chapter explores what liberation theology tells us with regards to how we might recognize our ecological sin and take up the journey of ecological conversion, with a particular emphasis on theologies of the Third World or based in the Global South. So the chapter discusses the work of theologians like James Cone and Leonardo Boff, as well as arguments from 'Third World' theologians such as Samuel Rayan, Tissa Balasuriya, Ivone Gebara, Aruna Gnanadason, Mercy Amba Oduyoye and Frank Chikane. It also draws attention to the arguments of Indigenous liberational thinkers such as Jione Havea to understand what is required of us as faith-filled persons, not least the importance of joining in the cry to give land back, to give oceans back. What we will see is the debt that discussions on ecological sin and ecological conversion owe to liberational theologies. We will also see that even if these discussions do not explicitly name race and/or the ecology directly, they provide a critique of those systems of domination and exploitation that have created the ecological crisis within which we now live, a critique that we saw so clearly in the description of the Filipino bishops' document in Chapter 1. My reading of these theologies – perhaps because I read them as a woman from the Third World – sees a strong thread of righteous rage linking these demands for freedom and justice for all. They are, using the phrase that James Cone deploys to describe Black theology, 'fighting spiritualities'. After all, what is it that liberation theology asks us to do but to open our hearts and our minds to what is around us, to become affected, and to insistently ask the question 'Why', for a moral awakening that is affected by rage at the evils that structure our status quo. Boff says it in this way, that the main thrust of the liberation theologies of the 1960s was ethical indignation on seeing the wretchedness of poverty. This situation was unacceptable and called for the 'true sacred anger of the prophets'.[2] Is this an anger we still have today? Certainly it is an anger that liberative thinking from Black, Third World and Indigenous theology does express in the moral awakening that they demand. It is important to remind the reader here, as I have argued elsewhere, that these theological demands are not about an engagement in a kind of Europhobia or anti-Westernness, but it is an embrace of an internationalist position, a recognition of the conjoined struggle that all of Creation must engage in to defeat unjust dominant structures. Liberational thinking critiques the Enlightenment values of the self and, as the documents referred to in the previous chapter also do, urges us to make central the notion of the multiple Other, even as all the while we long for the Profound Other.

'A conversion to liberation, an illogical conversion'

Liberation theology's defining characteristic is that it is a praxis that is lived and lived in solidarity with the poor and oppressed. It is a questioning posture of which freedom is the goal, and it requires three key moments: the moment of praxis, the moment of reflexion on praxis and the moment of a return to a renewed praxis. As Zoë Bennett notes, liberation theology begins and ends in praxis.[3] Gustavo Gutiérrez describes this as a theory of action, that human action must be understood as expressing an inner impulse and that this becomes real through what is expressed in action, the activity suggested by the relationship between faith and love.[4] Within this framework, praxis is activity that is in solidarity with the poor and the oppressed, and that contains within itself the absolute intention of bringing about a situation of liberation through a process of radical, transformative social and political change. Away with the songs and the sacrifices, what is offered up is the struggle. From this develops the question, how can poverty that is created by structural injustice be addressed? Liberation theology, broadly, is birthed in the experience of those who witnessed the suffering of those oppressed by various totalitarian regimes in the poorest nations of the world, particularly Latin America. Liberation theologians from the South take seriously the contexts from which theology emerges and begin their theological reflection with social analysis. Mario Aguilar notes that due to the agitations of liberation theologies, the materially poor entered into the mindset of the (Catholic) Church not to demand rights but to suggest a reality that challenged the rich and powerful life of the Church.[5] As Wati Longchar further notes, these very localized understandings of liberation and justice also offered a multiplicity of Christian selves.[6] This is a reference to feminist and womanist theology that reflects on the struggle of women, to Black theology that foregrounds the Black experience and the struggle of Black people, to the *minjung* theology that is a product of the struggle of the Korean people against the dictatorial regime of the 1970s, to the Dalit theology that focuses on the dismantling of caste discrimination, to the importance of a theology of suffering, and to those Indigenous theologies that demand attention be given to the complex spirituality of land and the inability to be free when the land is not free.[7] In liberational thinking, it is the experience, hardship and spirituality of those who have been historically marginalized that become the vital source for doing theology and imagining and engaging in the struggle for a world conceived otherwise. It is not only discerning God, as we will see below, but working collectively to transform unjust situations and provide a vision for the future. As such, we could argue, it comes from a space of generative rage.

The particularity of praxis within a context is also important and here I turn to what we see in the liberation theologies of the 'Third World'. Theologies of liberation have long made the connection between capitalist domination and ecological devastation. This understanding of the deep-rooted sinfulness of capitalism and plunder is underscored in the political theology that we find in Asia, particularly Asian theology that allies itself to Third World liberational thinking. This is a political theology that emerged from the struggle for political independence after the Second World War, as well as the critiques of colonialism and development. Many of the discussions here were significantly influenced by what we may term the 'Bandung moment'. The Bandung conference of 1955 brought together delegates from African and Asian countries in order to affirm their desires for independence as well as their agreed refusal to align themselves with the world powers of the First and Second World. Delegates were united in their opposition to colonialism and urged member countries that were still under colonial rule to fight for independence. The Bandung conference agreed on three basic tenets: the decolonization and emancipation of the peoples of Africa and Asia; peaceful co-existence; and economic development and non-interference in internal affairs.[8] Although, as historians of Bandung point out, there were different definitions and approaches among the delegates, the key thread of solidarity was a spirit of anti-imperialism in all its forms and the determination to be 'masters of their own fate'.[9] Lumumba-Kasongo also argues that the main focus of Bandung was to privilege a South–South agenda that wished to create a multipolar world.[10] This is a very essential entry point for understanding what Third World theologians then also pick up, especially in terms of building alliances from within liberational thinking and across the Latin American, African and Asian contexts.

Since Bandung, and then more formally in the 1970s, Southern theologians have decried the theological hegemony of Europe and North America and reclaimed the right to speak about God as subjects of their own destiny. As M. P. Joseph writes, the conference at Bandung was essentially an intellectual kick-starter for the gathering of the Ecumenical Association of Third World Theologians (EATWOT), noting the 'radical emotions' that Bandung wrought in the leaders and thinkers of Africa and Asia. Theologians wished to respond to this movement and to accompany it spiritually:

> Bandung had a profound influence on theological thinking around the world. As an acknowledgment of this influence, EATWOT was described as a 'Bandung of theology' by European theologian M. D. Chenu.[11]

From this Bandung moment, and through the 1960s, 70s and 80s, we see the terms 'third world' and 'liberation' beginning to be recognized as cognate concepts in theology. The idea of a 'third' world was not understood primarily as the number three in the numerical sense, following the first world of the capitalist countries and the second world of the socialist bloc. Rather, in accordance with its original French usage, it indicated a third something – a 'tiers monde' that connoted an alternative world to which the decolonized nations of the 1950s and 1960s aspired: a world different from either of the other two, something between the capitalist and communist systems.[12] The Non-Aligned Movement, an association of states that did not wish to formally seek alliance with either the United States or the USSR and whose formation many directly trace back to the Bandung conference, was the most expressive political symbol of this new awareness.[13]

The Christians of the Third World who appropriated this political consciousness increasingly rejected the development model of the 1960s and supported the liberation agenda of Third World leaders for whom the two 'developed' worlds were centres of domination. Historians of the EATWOT note that there were three key drivers for such theologizing. The first was the need to seek a solidarity that was informed by a search for subjectivity among those who were historically marginalized. The second was to reject European epistemology – including Western theology and Western church institutions – in both economic and political logic. Per Frostin describes this as a direct uptake of renewing theological methodology, pointing to Third World theologians' explicit rejection of academic theology divorced from praxis.[14] In particular, in the early stages of the EATWOT conferences, we see an emphasis on the importance of the dialectic between theory and praxis. This is to proclaim 'another way of knowing the truth' – acknowledging that the world is an unfinished project and asserting that knowledge occurs through immersion in the project and process of transformation.[15] Ivone Gebara's understanding of a praxis-oriented theology is instructive here. Gebara not only engaged with a re-reading of Scripture, she also critically engaged her socio-economic context and focused on deconstructing knowledge using a liberationist perspective.[16] In doing so, her focus was to understand silences, that is, to understand the hidden forms of oppression that had become normalized through patriarchy and development-based hierarchies. Her liberationist thinking was focused not only on understanding the needs of the marginalized, but also in discerning avenues of transformation. Elaine Nogueira-Godsey, narrating stories from Gebara's life, notes how her engagement with the daily realities of poor women made her understand the importance of in-between spaces, and the limits of liberationist thinking if it did not realize that domination was deeply rooted in people's anthropological and cosmological worldviews.[17] Similarly, the Circle of Concerned African Women Theologians, which began convening during the days of the EATWOT, sought to take seriously the experiences of culture, gender, class, ethnicity and race that African women encountered, critically evaluating culture and religion in order to expose the ingrained dominations of patriarchy, sexism and imperialism.[18] This was not only about women's struggle but also about the struggle for liberating all of humanity, and the importance of noting the ways in which domination takes root.[19] Mercy Amba Oduyoye provides a very practical example in her reflection on the African perspective,[20] noting that at the time of writing, the whole of Africa was in thrall to neo-colonial economies, and part of these economies was the type of Christianity being promoted by the West. Oduyoye calls out this Western Christianity for being allied to the wealth of multinationals that sought to entrap Africans so that 'they may serve as its labour force and [become] cannon fodder for its militarism'.[21]

The final driver was to work towards greater ecumenism and interreligion. In Asia, in particular, this was important for building a broad coalition of all religions together who would explore and share their 'liberative elements' and work together to overthrow oppressive systems, especially because through the 1970s and into the 90s many Asian nations were also going through particular turmoil, EATWOT theologians like the Sri Lankan Tissa Balasuriya and the Korean Prof. Soon K. Park were often under state surveillance and even faced imprisonment in the case of the latter. Fr Tissa, famously, was briefly excommunicated by the Catholic Church. As the EATWOT's first president R. Chandran proclaimed, Third World theology was the 'bitter fruit' of oppression, it 'comes alive in the totality of an oppressed people to be fully human'.[22] Joseph quotes the scholar Mveng's observations on these dialogues as well, where he notes that what binds these movements together is the fact that

> The liberation context requires constant mobilization against the forces of oppression and exploitation that trample down the weak, poor and the oppressed people, the post-liberation context inaugurates a dialogue between the gospel and Marxist socialism, in the form of religious freedom which is contextualized.[23]

As K. C. Abraham notes, this was a taking up of a liberational praxis.[24] Drawing attention to the Charter Identity of EATWOT drawn up in 1976, he describes the radical break here as epistemology that politically commits itself to action and engages in critical reflection on the Third World. As Joseph and Mario Aguilar both note, that reflection on praxis is the first act of doing liberative theology.[25] Such a liberational praxis also underscores why these Third World Christians did not hesitate in dialoguing with radical Marxist movements in the contexts of their country. Indeed, historians of the EATWOT note how, in 1987, due to the recognition of socialism as a passion for many EATWOT theologians, there were even organized dialogues between theological groups and countries that were undertaking socialist models, particularly China.

While engagement with Marxist movements was important, Third World theologians like Aloysius Pieris also cautioned against using Marx in his dogmatic form, noting that he was a European of his time who did not understand the seriousness of the religio-cultural realities that existed outside Europe.[26] Drawing from Pieris, as well as Cornel West's keynote speech to the EATWOT, Frank Chikane notes that Third World theology must blend Marxist tools of analysis with religio-cultural tools of analysis and commit to a radical transformation of society.[27] However, the tools for doing so are developed in the course of the struggle, they are 'tested in the praxis of struggle'. This underscores the importance of taking on a posture of question, investigation, encounter and mobilization, elements that we see in the ecological justice movements that form the empirical case of this EATWOT paper. These form the elements of such a struggle because of the importance of being affected by lived experience. As Abraham and Bernardette Mbuy-Beya discuss in *Spirituality of the Third World*,[28] the central critique here is to know, by engaging praxis and struggle, that the starting point was to be able to comprehend a God of life against a situation of the massive death of the poor. Third World theologians had to contend with the failure of the historic socialisms of the East as well as the development capitalisms that were prevalent in the South, such as that which Oduyoye notes above. Abraham and Mbuy-Beya remind us that in the Third World development capitalism presented itself with a human face, but that after the end of the Cold War and the imposition of what we might call the age of American imperialism (my addition), we now have a kind of savage capitalism.[29] This savage capitalism concentrates its power in the North and, utilizing logics of coloniality and race, unleashes its aggression on the South. As Mercedes Canas notes, reflecting on environmental degradation in El Salvador, this is a form of development that destroys vegetation, forests, mangroves, water and children in equal measure.[30] It sustains itself by destroying nature and excluding from consideration the lived experience of the majority world.

For Asian liberation theologians of the Third World, the fundamental starting point is the lived experience. Asian theologians of EATWOT identified that the primary agents of oppression in Asia were monopoly capitalism, landlordism, and the imperialism that was, by then, represented by multinational corporations. Asian theologians were and are insistent that this oppression occurred through foreign aid, loans, imported technology and military alliances.[31] It is important, therefore, that such theological work is both part of political struggles, of movements, and also works to create or argue for alternative economic relationships. For Asian theologians, work towards solving the issue of human and environmental degradation began in solidarity, through a joint involvement in common struggles.

The Asian context was seen to be defined by two realities. In the first, there is enforced poverty and oppression by the forces and structures listed above. The second reality is the liberating aspect of poverty, taking on a simpler life and therefore engaging with the struggle against Mammon.[32] It is quite interesting that for Asian theologians, especially those hailing from Dalit and Indigenous backgrounds, Jesus Christ's divinity has never been of prime importance; rather, they focus more rigorously on Jesus as the poor monk who wandered from village to village. Therefore, as Aloysius Pieris elaborates in his seminal text *An Asian Theology of Liberation*, there are tools of introspection or self-analysis proper to Asian religions that inculcate liberation from personal greed (the acquisitive instinct), as well as tools of social or class analysis that expose the mechanism of institutionalized greed (unbridled capitalism). All are employed in conjunction with each other as a strategy for a liberative praxis which, according to this theological space, is also the first formulation of a liberative theory. For Simon Kwan,[33] this provides the basis for how resistance is understood in Asian theology: as a discourse that seeks to resist the colonialism that has been constituted in socio-economic and socio-political exploitation, *as well as* the theological consequences of Western universalism. Indeed, Kwan, working with Hall's notions of identity and Bhabha's work on hybridity, sees a clear importance in what he calls the resistive identity of Asian theology which, via praxis, is anti-colonial without being binary, and is able to keep revisiting and reshaping the boundaries of its identity. These are calls to convert to resistive, refusing identities. This is not simply an analytical moment but, as we can see in what is asked by the theologians of the Third World, it is about a kind of rage that demands a moral awakening, for justice now and not at the end of the world. Tissa Balasuriya directly names this as requiring a moment of conversion.[34] In his work *Planetary Theology*, he argues that what is required is a conversion to liberation. This is engendered by a growing consciousness of domination and those elements within our current status quo that allow oppression and domination to occur. For Balasuriya, awareness goes beyond simply having information; it is a process of evaluation and motivation that engages the whole person. Balasuriya sees struggle as necessary and also cleansing. The oppressed must struggle, he says, to overturn domination, while the oppressors must undergo a 'self-purification' of their exploitative attitudes. This struggle is one in which the oppressed are widely mobilized to work for liberation, and thereby break down the divisions among dominated groups. This conversion towards the liberation struggle is about action, about transforming mentalities and social structures, and is a commitment that must be made over and over again.[35]

The Indonesian theologian Marianne Katoppo would concur. She uses a reflection on the Eucharist to make the same point.[36] Katoppo sees the Eucharist as a visible means of breaking down socio-economic barriers. Speaking of the Eucharist as a sign, she asks us to reflect upon it, and draws on Balasuriya's remark that Jesus Christ only ever celebrated the Eucharist once.[37] Not long after this he had been killed by the unjust social structures from which he was seeking to liberate his people. For Katoppo, this is what the Eucharist signifies, and this is certainly a radical epistemological break from Western thinking on the Eucharist as sign and symbol. Certainly, I would argue that this sees the Eucharist as 'sending us out' in the way that Francis does in *Laudato Si'*, as well as further underlining its praxis. Thinking about the Eucharist and placing it in the Asian context, Katoppo discusses that what is suggested here is a theology of the womb, one which understands salvation as the chief concern of religion. For Katoppo, salvation is struggle, that is, involvement in the struggle for peace and freedom. Like her Asian confrères, Katoppo notes the importance of people taking up the struggle in order to fulfil 'ourselves in the passage of time'. She links this to the theology of aching, as elucidated by the Vietnamese theologian C.-S. Song,[38] noting in particular his observation that

Creation, apprehended in the perspective of redemption is the true foundation of human community in which begins the kingdom of God, that links all humanity in a common kinship and blood relationship.[39]

An interrelatedness is also clearly present within the framework of a praxis-oriented thinking from our faith-based perspective. What does this mean when thinking with ecological discourse?

Resistance, refusal and bleeding wounds

Although, arguably, liberation theology is initially focused on human survival and flourishing, the ongoing association of marginalized communities, especially Indigenous communities, with the devastation of land, as well as the recognition of the link between enforced poverty and the struggle for land rights, means that the 'cry of the poor' must also be understood alongside the cry of the Earth. This is, in some ways, fully articulated in the 1995 work of Leonardo Boff.[40] How Boff draws together the strings of liberation theology, particularly Indigenous thinking and commentary on ecological justice, focuses us in this way and reminds us to pay attention to 'two bleeding wounds'. He argues that the first wound is one of poverty that breaks the social fabric of millions. The other wound is the systematic assault on the Earth; this breaks down the balance of the planet, which he argues is under threat from the plundering of development as practised by contemporary global societies. He then connects the reflection on these wounds to the 'cry' of the poor for life and freedom and beauty, and the Earth crying out against oppression. For Boff, the poor seek liberation as organized historical agents while the Earth seeks freedom through a new alliance with human beings. He further argues that liberation theology meets the ecological question when it reflects on the scale of the human tragedies caused by the climate crisis. Boff sees liberation theology as well placed to do this due to the inherent critique of development and capitalism that is already present. Certainly, we see this very strongly in the discussions above, especially those concerning Third World theology. Boff notes that the critique suggested by liberation theology is to denounce the economic developmentalism advanced by dependency theory and world economic systems. As such, the liberational concern for modernity is rooted in praxis. Poverty, therefore, is an environmental dilemma. The great crisis is catalysed by a 'culture of the satisfied', that is, those elites who benefit from the current economic system, as well as the desperate attempts of the poor to survive. A logic of sinfulness, Boff finds, drives dominant populations to oppress the marginalized and plunder the Earth, refusing the strong interconnection that binds the ecological, human, social and spiritual aspects of life. Boff argues that this sinfulness has deep roots in the spiritual malaise that characterizes capitalism. Indeed, in something of a similar way to that in which Leon Sealey-Huggins warns that exclusively technical solutions are not enough,[41] Boff and other liberation theologians argue that what we require is a movement away from seeing logic in a ruling system based on profit and social manipulation. To address this, we require something more than social justice, and it is here that we see the strong link between Boff's thinking and the ecological conversion of Francis. Boff sees the building of a new alliance between humankind and other beings, a kind of cosmic community, and argues that liberation theology must now adopt the new cosmology of ecological discourse.[42] There is a need to make a moral choice to move away, collectively, from unfettered profit and growth to building a 'logic of the common good'.

Ivone Gebara's work takes up thinking on the ecological crisis and also sees poverty as part of this crisis and calls our attention to those that she sees as the victims of unfettered growth: Indigenous peoples, women, rainforests and natural biodiversity.[43] She decries arguments within feminist spaces that wonder at the primacy of ecology over feminism, arguing that while we attempt to settle these disagreements, the world is struggling against deforestation, pollution and starvation brought on by a capitalist system that secures profit and life only for a chosen few. In this way, Gebara's argument mirrors Boff's. Gebara argues very strongly that there is a need to go beyond competition in order to make another world possible, and that the solution cannot come from a patriarchal system of dependence.[44] This structure reproduces violence and such violence is suffered by the Earth as well as by the marginalized. Boff speaks of a need to move collectively towards a common good, and Gebara, in parallel, asks for a new national and international order that brings together civil society to structure a qualitatively new daily life.

James Cone's writing on liberational thinking and the ecological crisis is also concerned with the centrality of a moral stance. In his writing on liberational theologies, as well as his essay 'Whose Earth is it anyway?',[45] Cone writes simply but beautifully of the necessary relationship between faith and praxis. Cone notes that our Christian faith tells us that God created us all for freedom, and this therefore means that we engage in praxis as social theory that we use to analyse social structures that are unjust. This analysis helps us to discover what must be done to realize the freedom to which all are entitled. Cone notes that our faith is not only our love for God, but a love for God that is expressed through love for our neighbour. Such love demands justice as the act of loving because the Divine calling to us 'requires more than a pious feeling'; it demands social and political analysis so that we escape the trap of substituting piety for justice.[46] In 'Whose Earth is it anyway?', Cone also makes explicit the link between the struggle for racial justice and the struggle for ecological justice. He describes the 1991 gathering of the First National People of Colour Environmental Leadership Summit, bringing together grassroots and national leaders from the Americas, all of whom made clear the fact that people of colour were disproportionately affected by environmental pollution. Cone calls for the connection to be made between racism and the degradation of the Earth as a much needed work, and quotes from the work of womanist theologians such as Emilie Townes, Delores Williams and Karen Baker-Fletcher, who were already noting the parallel defilement of land and bodies of colour, especially female bodies of colour. Cone notes that the same logic that structured slavery and segregation is also the logic that destroys and strips away the Earth. For him, struggles for ecological justice that do not also fight white supremacy are racist, and those that do not include ecological justice in their work for racial justice are anti-ecological.[47] Cone, much as Pope Francis would do later in *Laudato Si'*, uses the strong language of sin and evil to make these connections, stating that

> Racism is profoundly interrelated with other evils, including the degradation of the earth.[48]

These interrelated evils must be encountered and overcome together, and in thinking of this interrelationship, we see once again the same argument that was present in our previous chapter when it laid out the terms of ecological conversion and how it was linked to an integral ecology.

The Indian Jesuit Samuel Rayan makes a similar argument, but with a different approach: he focuses us on the question of right relationship and cosmology.[49] Indeed Rayan, as well as Aruna Gnanadason and Gabriele Dietrich (discussed below), makes a very deliberate shift away from the theological anthropology that we see in Boff's work. Rayan provides a narrative of how the Fijian islanders were taken aback by the first Christian missionaries, not only because of the layers of clothing they wore but because the religion they preached had little to do with the land or the sea. Rayan goes on to argue that the Divine that dwells in the heart of reality is the central message of all cosmic religions, and we see this in the Upanishads and even in the Judeo-Christian psalms, because the whole universe is pervaded by the Divine. All anthropocentrism is out of place here. According to Rayan, turning away from this understanding is sinful, because he argues that an anthropocentric theology insulates us humans from our 'rootedness in the cosmos'. Rayan asserts the need to understand that the Creation became under the Spirit of the Divine which, in its brooding over the waters, turned chaos into the cosmos. Thus Rayan foregrounds the contemplative traditions of Asia in order to make an argument for the aspiration for freedom and the readiness for struggle, or what he terms a response-ability. What is required of us is a critical openness that understands the realities of nature and history, that then engenders a willingness to act to transform reality. In something of a similar vein to Deane-Drummond's discussion of 'deep incarnation', Rayan discusses the New Testament proclamations of John 3.16, 2 Corinthians 5.19, Romans 8.21 and Ephesians 1.10 to highlight the fact that the story of salvation is told through Christ in the 'memory of creation', wherein all of the heavens are evidence that God gives of themselves.[50] All must be seen in the context of an ongoing, dynamic Creation.

> Creation is God's family born of his heart, and loved by her into existence and nurtured daily with endless caring. Of this cosmic family of God we are (to be) response-able members.[51]

In all religious discourse, Creation reveals the Divine and so it is there to be contemplated and seen as part of creative activity. Our turning away from this understanding is sinful, and we must realize our ecological sin.

This revelation of the Divine, as well as the interrelatedness of the whole inhabited world, is important for the work of Aruna Gnanadason. Gnanadason notes that there is a first truth we must encounter, an integrity.[52] In a similar fashion to Rayan's discussion of the rootedness of Creation, Gnanadason sees the integrity of Creation as calling us to work ecumenically to help all life thrive together. Critiquing liberation theology's anthropocentrism, in a similar way to Rayan, Gnanadason reminds us that we must also foreground the Earth's struggle for integrity. That is, refuse an anthropocentric approach to God and theological anthropology, but remember the violence that the Earth as God's own body is often subjected to.[53] Gnanadason, in her descriptions of a feminine paradigm for the ecology, further notes how women in South Asia frequently understand the link between the suffering of the Earth and the violence that women endure.⁵⁴ Like Rayan, she cites the connectedness of Indian cosmology, which sees person and nature as having a duality in unity. So the struggle that is taken up is the resisting of death.

> Resist the digging of mountains
> That bring death to our forests, streams
> A fight for life has begun.
> (Chamundeyi, Doon Valley)⁵⁵

As noted in the section on Cone above, this idea of seeing environmental violence and violence to women as interlinked is also something we see in the work of womanist theologians like Delores Williams. In her essay 'Sin, Nature and Black Women's Bodies', Williams decries the logic of domination, insisting that we must work for planetary freedom as well as bodily freedom for women.⁵⁶ Williams discusses the enslavement of African women in order to underline that the domination logic that sits at the heart of modernity is just as lethal to Black female bodies as it is to the Earth. Her key example is the impact of strip mining on the reproductive capacity of women and the rape of Black women for breeding slaves. For Williams, just as it is sinful to defile and rape the body of the Black woman, so too it is sinful to defile and rape the body of the Earth.⁵⁷ Dietrich also makes this argument regarding the need to see the Earth as the body of God, and that realizing this places upon us the impetus of working towards equity.⁵⁸ An understanding that the 'Earth is the Lord's' requires a radical critique of property rights, an abolishing of slavery and also opposing capitalistic organizing that privatizes the land, the sea, the rivers and even drinking water. We must ensure that environmental protection is not divorced from people's survival rights. However, such work requires intense power struggles, a revolutionary Kairos. Dietrich argues that we must see the world in the dialectic that occurs between the organization of society and human community and nature, a body that is a whole. In other words, we must oppose fragmentation.

Land back, oceans back

It is fundamental to all this talk of liberational thinking to discuss what the voices of Indigenous theology are also arguing for. Any discussion of liberation that does not place Indigenous voices at the centre will be incomplete. Indigenous voices are often silenced in consideration for global justice because frameworks for justice often only seek spaces in which understandable ideas of justice are already present. It is because of this silence that Indigenous scholars like Eve Tuck and Leanne Betasamosake Simpson speak of refusal. Betasamosake Simpson reminds us that this is the refusal of colonialism in order to affirm different worlds in time and space.[59] Krushil Watene argues that Indigenous peoples are often silent in justice theorizing and, as we have seen in the various COP gatherings of the past decade, are denied an active role in conceptual and policy landscapes.[60] Indigenous communities are globally present, and are understood as those who have a historical continuity with pre-invasion and pre-colonial societies that developed in their native territories. Throughout the world it is the tribal and Indigenous people who have been significantly marginalized and oppressed, whether it is the First Nations communities of Canada or the Veddas, the forest dwellers, of Sri Lanka. Several Indigenous communities that have been Christianized also express their theology from this space of suffering and oppression. This is a theology that attempts to express the Christian faith from the point of view of these experiences of hardship, and through the liturgical and traditional thought of Indigenous peoples. For Wati Longchar it is a resistance theology spoken or written by those who have become environmental prisoners on their own land.[61] Longchar elucidates that where Indigenous theology departs from 'mainstream' liberation theology is in the seeking of liberation from the perspective of space. As the Supreme Being is conceived of in terms of Creation, the issue of space is a central issue of justice. Dina Cebrian sees the struggles of resistance within Indigenous theology as a space from which strength is drawn to resist oppression and to dream.⁶² Speaking from her particular Andes context, Cebrian notes that the root source of her theology is the revelation of the God of life in Indigenous culture and history, which she argues pre-dates the introduction of Christianity. As such, the subject of theology is the community itself, the community that constantly seeks life and does so in cooperation with the God of life. With this, the major source of Indigenous theology for Cebrian is the concrete action taken to care for and defend the life present in the trees, the forests, the water. Jione Havea, the Pasifika theologian, in his discussion of theologies of the sea, speaks with keen emotion of the troubling of the waters, placing this experience of hardship at the centre of his critique, together with the need for the expression of resistance. He describes how European settlers enter their civilization into Pasifika waters by bringing with them the diseases of smallpox, typhoid, syphilis and gonorrhoea. The colonizer 'infected native bodies with diseases ... and colonised the minds of generations to come ... they darkened the ways of our ancestors'.⁶³ As such, Havea describes the work of Pasifika theology now as freeing theologies within this context from these physical and mental infections.

Havea speaks of a theology of the sea. Other Indigenous theologians argue for a theology of the land. For these theologians, the central argument is the special relationship with the land. Here land forms the basis of the entire social and economic system. If we think back to the story of the Fijian islanders that Samuel Rayan narrated, we will remember how the islanders could not fathom a God who had no connection to the land, because the land is what provided life. Wati Longchar notes that there needs to be an understanding that for many Indigenous peoples land carries a complex spiritual component and is not only sacred but is seen as a co-creator with the Creator-God. The land provides identity and is also the conduit through which a people becomes one with the Sacred Power. Sifiso Mpofu, citing the case of Zimbabwe, notes that the alienation of Indigenous communities from their land has brought about a situation of social evil.⁶⁴ As such, the Indigenous peoples in Zimbabwe based their liberation struggle on the need for land. For Mpofu what this means is that liberational thinking must develop a theology of the land in order to address the discontent and suffering faced by the historically marginalized. This kind of *theology as resistance* would immediately address the anger of the disenfranchised and also the emptiness of industrialized societies, while reducing to nothingness the ongoing presence of empire. Jude Fernando, discussing landlessness, wonders if the theological imagination can come to understand the yearning for land and the strength of community bonds linked to land.[65] In order to build such a theology, Fernando argues for adopting hermeneutical tools of land that envision a future beyond the relationships of empire, and focusing on those spaces where we have observed sites of concrete resistance. This is a theology that also takes seriously the distinct identities developed by native peoples. A strong example of this can be seen in Ferdinand Anno's description of the context in which the Indigenous peoples of the Gran Cordillera in the Phillipines 'do' theology.[66] Anno notes how, against the rapacious development projects of the twentieth century, the *pattong* dance was crafted to tell the tale of anti-colonial resistance to development aggression and to showcase the counter-discourse that interwove the ecological with the anthropological and by doing so underlined the importance of the human collective as it reflects the truly divine ordained cosmic community. Thinking about this centrality of land, Longchar argues that from the perspective of making land central, Indigenous theology focuses on the solidarity of collective resistance for the defence of the oppressed. This is a weapon through which the oppressed can challenge the system. According to Longchar, Indigenous theologians would seek through collective resistance to promote values and structure that enhance life and liberate from bondage.[67]

Replacement, awakening

But liberation, as the Sioux scholar Vine Deloria reminds us, is also a process of rejection or refusal.[68] That is, rejecting what we have learnt and replacing these things with what our lived experience tells us does have value. It is not only making central the land or the sea or the suffering of the marginalized but also rearranging how we perceive the world. This liberation begins in the destruction of the entire complex of Western or European or Enlightenment understanding, the radical break asked for by Third World theologians. After this will come the building of a new system that brings together a comprehensive understanding of human knowledge and experience. This rearrangement or refusal is the first step towards liberation. Refusal here is a generative practice, as mentioned in the introduction to this book. It is not simply a process of saying no; it is the process of rejecting narratives that dehumanize or humiliate and instead building narratives that involve histories, allegiances and a collective future.[69] As theologians we must be guided by indigenous resistance and the refusal of colonialism. Indigenous scholars note how Indigenous nations and their life determinedly build a different world that foregrounds the material and spiritual needs of the community. Here, the focus is on creating a world of deep relationality that is both local and international and involves complex forms of solidarity that share land and space in order to bring forth more life.[70] To do this requires moral commitment.

The journey to a liberated Earth can only happen if there is a moral awakening. More importantly, it must start in transformation and surrender to a collective struggle for liberation that works towards a cosmic and planetary community, and that reclaims the narrative from 'encroachers and manipulators'. What all these voices point us to is the breakdown of rightful relationship that Pope Francis termed the existence of ecological sin. The ecological crisis is the rupture of the communion between creatures and Creation. Theological voices from the Global South, from Indigenous communities and from liberational spaces have been telling us this for many years via their critiques of whiteness, capitalism, colonialism and other dominating structures. These critiques, even when they speak of journeys of love and hope, are not provided to us in a passive way, but in aching, in agony. They demand an awakening. One way in which we must journey to this awakening is to understand the histories and presents of exploitation and extraction. This will be the focus of the next chapter.

Notes

1 Sherry E. Jordan, 2015, 'Chung, Hyun Kyung, Interview, Ecumenical Decade of the Churches in Solidarity with Women Program', *Duke University Libraries*, 9 November, https://idn.duke.edu/ark:/87924/r4 wd3s44n, accessed 10.06.2022.

2 Leonardo Boff, 1995a, 'Liberation theology and ecology: Alternative, confrontation or complementarity?', in L. Boff and V. Elizondo (eds), *Ecology and Poverty*, Maryknoll, NY: Orbis Books, pp. 67–77.

3 Zoë Bennett, 2007, '"Action Is the Life of All": the Praxis-Based Epistemology of Liberation Theology', in Christopher Rowland (ed.), *The Cambridge Companion to Liberation Theology*, 2nd edn, Cambridge: Cambridge University Press, pp. 39–54.

4 Gustavo Gutiérrez and Robert R. Barr, 2004, *The Power of the Poor in History*, Eugene, OR: Wipf and Stock; and also Gustavo Gutiérrez, 1988, *A Theology of Liberation: History, Politics, and Salvation*, Maryknoll, NY: Orbis Books.

5 Mario I. Aguilar, 2009, *Theology, Liberation and Genocide: A Theology of the Periphery*, Reclaiming Liberation Theology, London: SCM Press.

6 Wati Longchar, 2013, 'Liberation Theology and Indigenous People', in Thia Cooper (ed.), *The Reemergence of Liberation Theologies: Models for the Twenty-First Century*, New York: Palgrave Macmillan, pp. 111–21, https://doi.org/10.1057/9781137311825_13, accessed 10.06.2022.

7 Longchar, 'Liberation Theology and Indigenous People', p. 115.

8 Christopher J. Lee, 2010, *Making a World after Empire: The Bandung Moment and Its Political Afterlives*, Global and Comparative Studies Series, no. 11, Athens, OH: Ohio University Press.

9 See Seng Tan and Amitav Acharya (eds), 2008, *Bandung Revisited: The Legacy of the 1955 Asian-African Conference for International Order*, Singapore: NUS Press.

10 Tukumbi Lumumba-Kasongo, 'Rethinking the Bandung Conference in an Era of "Unipolar Liberal Globalization" and Movements toward a "Multipolar Politics"', *Bandung: Journal of the Global South* 2, no. 1 (2015), pp. 1–17, https://doi.org/10.1186/s40728-014-0012-4, accessed 10.06.2022.

11 M. P. Joseph, 2015, *Theologies of the Non-Person: The Formative Years of EATWOT*, Christianities of the World, New York: Palgrave Macmillan, p. 26.

12 Volker Küster, 2010, *A Protestant Theology of Passion: Korean Minjung Theology Revisited*, BRILL, https://doi.org/10.1163/ej.978900 4175235.i-203, accessed 10.06.2022.

13 S. I. Keethaponcalan, 'Reshaping the Non-Aligned Movement: Challenges and Vision', *Bandung: Journal of the Global South* 3, no. 1 (2016), pp. 1–14, https://doi.org/10.1186/s40728-016-0032-3, accessed 10.06.2022.

14 Per Frostin, 'The Hermeneutics of the Poor – The Epistemological "Break" in Third World Theologies', *Studia Theologica – Nordic Journal of Theology* 39, no. 1 (1985), pp. 127–50, https://doi.org/10.1080/00393388508600037, acccessed 10.06.2022.

15 Sergio Torres and Ecumenical Dialogue of Third World Theologians, 1978, *The Emergent Gospel: Theology From the Developing World: Papers From the Ecumenical Dialogue of Third World Theologians, Dar Es Salaam, August 5–12, 1976*, Sergio Torres and Virginia Fabella (eds), London: Geoffrey Chapman.

16 Ivone Gebara, 'Feminist Theology in Latin America: A Theology without Recognition', *Feminist Theology* 16, no. 3 (2008), pp. 324–31, https://doi.org/10.1177/0966735008091397, accessed 10.06.2022.

17 Elaine Nogueira-Godsey, 'A history of resistance: Ivone Gebara's transformative feminist liberation theology', *Journal for the Study of Religion* 26, no. 2 (2013), pp. 90–106.

18 Helen A. Labeodan, 'Revisiting the legacy of the Circle of Concerned African Women Theologians today: A lesson in strength and perseverance', *Verbum et Ecclesia* 37, no. 2 (2016).

19 T. Hinga, 1996, 'Between colonialism and inculturation: Feminist theologies in Africa', in E. S. Fiorenza and M. S. Copeland (eds), *Feminist Theology in Different Contexts*, London: SCM Press, pp. 26–9.

20 M. A. Oduyoye, 2004, 'Commonalities: An African Perspective', in K. C. Abraham (ed.), for the Executive Committee of EATWOT, 2004, *Third World Theologies: Commonalities and Divergences: Papers and Reflections from the Second General Assembly of the Ecumenical Association of Third World Theologians, December 1986, Oaxtepec, Mexico*, Eugene, OR: Wipf and Stock.

21 Oduyoye, 'Commonalities', p. 101.

22 Ecumenical Association of Third World Theologians, Virginia Fabella and Sergio Torres (eds), 1983, *Irruption of the Third World: Challenge to Theology: Papers from the Fifth International Conference of the Ecumenical Association of Third World Theologians, August 17–29, 1981, New Delhi, India*, Maryknoll, NY: Orbis Books.

23 Joseph, *Theologies of the Non-Person*, p. 160.

24 K. C. Abraham (ed.), for the Executive Committee of EATWOT, 2004, *Third World Theologies: Commonalities and Divergences: Papers and Reflections from the Second General Assembly of the Ecumenical Association of Third World Theologians, December 1986, Oaxtepec, Mexico*, Eugene, OR: Wipf and Stock.

25 See Joseph, *Theologies of the Non-Person* and Mario I. Aguilar, 2021, *After Pestilence: An Interreligious Theology of the Poor*, London: SCM Press.

26 Aloysius Pieris, 1988, *An Asian Theology of Liberation*, Edinburgh: T&T Clark.

27 Abraham, *Commonalities and Divergences*.

28 K. C. Abraham and Bernadette Mbuy-Beya (eds), 2005, *Spirituality of the Third World: A Cry for Life: Papers and Reflections from the Third General Assembly of the Ecumenical Association of Third World Theologians, January 1992, Nairobi, Kenya*, Maryknoll, NY: Orbis Books.

29 Abraham and Mbuy-Beya, *Spirituality of the Third World*, p. 93.

30 Mercedes Canas, 'In Us Life Grows: An Ecofeminist point of view', in Rosemary Radford Ruether (ed.), 1996, *Women Healing Earth: Third World Women on Ecology, Feminism, and Religion*, Maryknoll, NY: Orbis Books, pp. 24–8.

31 Muriel Orevillo-Montenegro, 2010, *Jesus of Asian Women*, Moscow, ID: Logos Press.

32 Joseph, *Theologies of the Non-Person*.

33 Simon Shui-Man Kwan, 2014, *Postcolonial Resistance and Asian Theology*, New York: Routledge, https://www.taylorfrancis.com/books/e/9781134702541, accessed 10.06.2022.

34 Tissa Balasuriya, 1984, *Planetary Theology*, Maryknoll, NY: Orbis Books.

35 Balasuriya, *Planetary Theology*, p. 207.

36 Marianne Katoppo, 1979, *Compassionate and Free: An Asian Woman's Theology*, The Risk Book Series No. 6, Geneva: World Council of Churches.

37 Katoppo, *Compassionate and Free*, p. 82.

38 Choan-Seng Song, 2005, *Theology from the Womb of Asia*, Eugene, OR: Wipf and Stock.

39 Song, cited in Katoppo, *Compassionate and Free*, p. 83.

40 Leonardo Boff, 1995b, *Ecology & Liberation: A New Paradigm*, Ecology and Justice Series, Maryknoll, NY: Orbis Books.

41 Leon Sealey-Huggins, '"1.5°C to Stay Alive": Climate Change, Imperialism and Justice for the Caribbean', *Third World Quarterly* 38, no. 11 (2017), pp. 2444–63, https://doi.org/10.1080/01436597.2017.1368013, accessed 10.06.2022.

42 Boff, 'Liberation theology and ecology', p. 75.

43 Stephen B. Scharper, 1998, *Redeeming the Time: A Political Theology of the Environment*, New York: Continuum.

44 Ivone Gebara, 'ECOFEMINISM: A Latin American Perspective', *CrossCurrents* 53, no. 1 (2003), pp. 93–103. http://www.jstor.org/stable/24461123.

45 James Cone, 2020, 'Whose Earth is it anyway?', in Terence Ball, Richard Dagger and Daniel L. O'Neill (eds), *Ideals and Ideologies: A Reader*, 11th edn, New York: Routledge, chapter 9.69.

46 James Cone, 1981, 'Christian Faith and Political Praxis', in Brian Mahan and L. Dale Richesin (eds), *The Challenge of Liberation Theology: A First-World Response*, Maryknoll, NY: Orbis Books, p. 61.

47 Marguerite Spencer, 'Environmental Racism and Black Theology: James H. Cone Instructs Us on Whiteness', *University of St Thomas Law Journal* 5 (2008), p. 288.

48 Cone, 'Whose Earth is it anyway?'

49 Samuel Rayan, 1994, 'Theological Perspectives on the Ecological Crisis', in R. S. Sugirtharajah (ed.), *Frontiers in Asian Christian Theology: Emerging Trends*, Maryknoll, NY: Orbis Books, pp. 221–35.

50 Rayan, 'Theological Perspectives', p. 228.

51 Rayan, 'Theological Perspectives', p. 229c.

52 Aruna Gnanadason, 2005, 'Yes, Creator God, Transform the Earth! The Earth as God's Body in an Age of Environmental Violence', *Ecumenical Review* 57.

53 Gnanadason, 'Yes, Creator God'.

54 Aruna Gnanadason, 1996, 'Towards a feminist eco-theology for India', in Rosemary Radford Ruether (ed.), *Women Healing Earth: Third World Women on Ecology, Feminism, and Religion*, Ecology and Justice, Maryknoll, NY: Orbis Books, pp. 74–81.

55 Cited in Gnanadason, 'Towards a feminist eco-theology for India', p. 80.

56 Delores Williams, 1995, 'Sin, Nature, and Black Women's Bodies', in Carol J. Adams (ed.), *Ecofeminism and the Sacred*, New York: Continuum, pp. 24–9.

57 Melanie L. Harris, 'Reshaping the Ear: Honorable Listening and Study of Ecowomanist and Ecofeminist Scholarship for Feminist Discourse',

Journal of Feminist Studies in Religion 33, no. 2 (2017), pp. 158–62, https://www.opendemocracy.net/en/transformation/is-pope-francis-ecofeminist/

58 Gabriele Dietrich, 1996, 'The World as the Body of God', in Rosemary Radford Ruether (ed.), *Women Healing Earth: Third World Women on Ecology, Feminism, and Religion*, Ecology and Justice, Maryknoll, NY: Orbis Books, pp. 82–98.

59 L. Betasamosake Simpson, 2020, *As We Have Always Done: Indigenous Freedom through Radical Resistance*, 1st edn, Minneapolis, MN: University of Minnesota Press.

60 Krushil Watene, 2020, 'Transforming Global Justice Theorizing: Indigenous Philosophies', in Thom Brooks (ed.), *The Oxford Handbook of Global Justice*, Oxford: Oxford University Press, pp. 162–80.

61 A. Wati Longchar, 2012, *Returning to Mother Earth: Theology, Christian Witness and Theological Education: An Indigenous Perspective*, Tainan: PTCA/SCEPTRE.

62 Dina Ludeña Cebrián, 'The Sources and Resources of Our Indigenous Theology', *The Ecumenical Review* 62, no 4 (2010), pp. 361–70. https://doi.org/10.1111/j.1758-6623.2010.00076.x.

63 Jione Havea (ed.), 2021, *Theologies from the Pacific*, Postcolonialism and Religions, Cham: Palgrave Macmillan, pp. 3–4.

64 Sifiso Mpofu, 2019, 'A Theology of Land and Its Covenant Responsibility', in Jione Havea (ed.), *People and Land: Decolonizing Theologies*, Theology in the Age of Empire, Lanham: Lexington Books/Fortress Academic, chapter 6.

65 Jude Fernando, 2019, 'People, Land and Empire in Asia', in Jione Havea (ed.), *People and Land: Decolonizing Theologies*, Theology in the Age of Empire, Lanham: Lexington Books/Fortress Academic, chapter 10.

66 Ferdinand Anno, 'Indigenous Theology: Sources and Resources Perspectives from the Philippines', *The Ecumenical Review* 62, no. 4 (2010), pp. 371–8.

67 Gebara, 'ECOFEMINISM'.

68 Vine Deloria, Jr, 'A Native American Perspective on Liberation', *Occasional Bulletin of Missionary Research* 1, no. 3 (July 1977), pp. 15–17, https://doi.org/10.1177/239693937700100303.

69 Eve Tuck and K. Wayne Yang, 'Unbecoming Claims: Pedagogies of Refusal in Qualitative Research', *Qualitative Inquiry* 20, no. 6 (July 2014), pp. 811–18, https://doi.org/10.1177/1077800414530265.

70 Betasamosake Simpson, *As We Have Always Done*.

3

'The Plantation is Always With Us': The Histories and Presents of Extraction and Domination[1]

[T]he events of today's changing climate, in that they represent the totality of human actions over time, represent also the terminus of history. For if the entirety of our past is contained within the present, then temporality itself is drained of significance.[2]

The arguments in this chapter are not new ones. The facts relayed in this chapter are not, in any way, a kind of original scholarship, and I do not make any claim towards this. It is the retelling of histories and presents that must trouble and affect our theologizing and our existence as persons of faith. It is the retelling of acts of deep sin in which we have all been complicit and that continue to separate us from the Creator. It is to discuss the 'civilized savagery' of colonial logics and how they are replicated in the international system of the present.[3] Colonial powers were, as Frantz Fanon argued, dependent on the Othering of those that they colonized, relying on brutal and barbaric violence – both physical and psychological – in order to create their industrial order.[4] This is not a condemnation of the European actors only, but also of the national and local elites who colluded with the colonizers to oppress those from marginalized backgrounds. I cannot but quote from Aimé Césaire, from one of his most famous pieces of writing that Robin D. G. Kelley refers to as almost a kind of Third World manifesto,[5]

Whether one likes it or not, the bourgeoisie, as a class, is condemned to take responsibility for all the barbarism of history, the tortures of the Middle Ages and the Inquisition, warmongering and the appeal to the raison d'Etat, racism and slavery, in short everything against which it protested in unforgettable terms at the time when, as the attacking class, it was the incarnation of human progress.[6]

What we know and what we want to know

This chapter is concerned with describing the long history, as well as the present nature, of international extraction. It does so in order to underline how colonial logics still operate and why activists and scholars call our attention to this in order to insist on the need to transform the world order. This is to say, as hopefully this book consistently notes, that we cannot engage in working for ecological justice if we are not simultaneously engaged in anti-imperialist and anti-racist endeavours.[7] This will take much unlearning and re-learning. In order to undergo this process, we must be aware of what has occurred in history, and be willing to listen to those peoples and voices that demand our attention for marginalized histories. In speaking of this process of unlearning, I speak also from my own experience. On one side of my family there were many wealthy uncles and grand-uncles who were often referred to with deep respect as 'planters'. It was only in reading about and researching the history of tea and rubber plantations in Sri Lanka that I found out, to my horror, how their wealth had come about.[8]

An important aspect of recognizing and repenting of our ecological sinfulness is to learn this history and to see how it is replicated in the present. This process of learning is also important for fuelling and making robust the claims and actions we must make in prophetic rage. We cannot remain ignorant but must recognize that our moral understanding of the world is based on epistemic and political structures that make it difficult to fully acknowledge the long histories of discrimination and international exploitation.[9] As James Baldwin has thundered, we do not know it and do not want to know it.[10] This is not simply a lack of knowledge; it is the difficulty of recognizing realities when we live in a system of individual interaction with multiple factors within society. This recognition, as well as a process of rejecting that which is known as being the only kind of knowledge, is a key step towards justice.

The sociologist and public pedagogue Gurminder Bhambra agrees, arguing that there is a need to understand how the world is configured so that we can begin to address inequality and injustice.[11] To reject what we know, we must first understand how and why we know it, and then lay bare these histories as they shape the present. That is, to understand the intimate relationships between power and knowledge and then to refuse what we know and move towards epistemic justice. This is to understand that there are power asymmetries in both knowledge production and reproduction. This is not exclusive to the dominance of Western perspectives and concepts, but also about blind spots or processes of silencing in knowledge production. The Haitian anthropologist Michel-Rolph Trouillot, who wrote very specifically on how knowledge production imposes and creates silences, also noted the need to make visible the cultures, histories and powers that colour epistemic processes.[12] Trouillot argued that these silences are found not just in academic histories, but also in sources, archives, the ways in which societies remember the past, tell stories and establish historical significance.[13] Certainly, this is in parallel with the arguments of Edward Said, who taught us that human self-knowledge is constantly being made and unmade, and that there is a need to always rethink and reformulate historical experiences.[14] This refers most particularly to the processes through which historical distinctiveness is forged so that it is both unknown and yet knowable, present and yet unrecognized. Building on all of this thinking, Santushi Amarasuriya et al., looking at particular acts of resistance and refusal, underline both Said's and Trouillot's work to argue for paying attention to refusal in terms of social and cultural relations, as part of the conditions of possibility for such refusal.

> Theology has to review past conceptions, see if others are needed, and in light of new problems, update old ways of looking at things that fit only the experience of the past and not the major questions of the moment.[15]

Although much of the work I have referenced above is from the social sciences, within liberation, Black, Feminist and Indigenous theology there has been a long anxiety regarding the need for epistemic justice.[16] The theologians of the Third World that were discussed in Chapter 2, for example, intensely agitated for an epistemological break from the West. Gutiérrez's argument in *The Truth Shall Make you Free* is that epistemic and spiritual injustices are mutually reinforcing.[17] For Gutiérrez, theological thought and theological praxis must make epistemic justice their central aim. Similarly, Dalit theologians noted the importance of understanding historical and sociological realities as well as theology that is committed and praxis-oriented. Let us not forget what was discussed in Chapter 2, that the theologians of the Global South explicitly rejected theology that was divorced from reality and from action. The radical break they propose is what makes the first act of theology one of commitment and is oriented towards critical reflection on the reality of the Global South.[18] Picking up this critical reflection is also to appreciate and recognize that alternative ways of knowing and being have persisted in the living traditions of those who have been colonized. As Chapter 2 showed, this is particularly true when discussing Indigenous theology. Robbie Shilliam calls this the tenacious vitality that reminds us always of thought and action that is constantly oriented towards global justice.[19] Theology's purpose is to transform and a radical approach of epistemic justice is fundamental here, one that is full of a theological imagination that embraces the pluriverse, that 'knows otherwise' than the universalism of colonial logics and resists them. This is a liberating action, an action of love. An important part of making this break is also the awareness of what is within history. So what do we need to recognize? First is the enduring nature of colonial logics, and second is the fact that the history of the disposable human and the disposable ecology are bound up in one another. To privilege the one over the other, or to fetishize the non-human over the human or vice versa, is to work within the confines of an incomplete vision of justice. Extractive patterns of accumulation are an enduring colonial logic that must be recognized and transformed as part of the struggle for ecological justice.

Histories and presents

There will be a few 'cases' presented within this chapter. Their purpose is simple: to underline that histories and presents are complex things and are completely intertwined with each other. The long histories of extraction that have brought us to the ecological crisis stem from colonial logics that continue to structure and moralize the present. So how did we get to here? It is a fact that the modern, capitalist world economy is one in which colonizers and colonized were systematically bound into relationships of extraction, colonization and dispossession. It is also important to note here that this is not simply a tale of European colonization, but also of the collusion of elite and upper-class/caste persons in the colonies with the colonizer to deem certain lands and certain communities as lesser and disposable. As Sankaran Krishna notes, in his excellent study of the struggle of the Dongaria Kondh people against the Vedanta mining company, the primary beneficiaries of development were, and still remain, upper-caste/class urban communities.[20] Those not falling into these categories were displaced and made destitute. Sir Roger Casement, looking at the case of the Peruvian Amazon company in Putumayo, remarks that once an indigenous tribe was taken over it was made the exclusive property of its conqueror.[21] Local magistrates, he noted, 'actively intervened' to capture indigenous escapees. As such, this is not only a question of what the West did to the East, but also how Eastern elites profited from the colonial era. In a recent study I did on race for Christian Aid, a key discussion point was how narratives of purity and superiority that were rooted in the Brahmin caste system that dominates much of South Asia were reinforced through imperial culture.[22] The socially constructed nature of these colonial and racist ideologies reinforces deeply unequal systems which position people with darker complexions or from minoritized backgrounds as inferior. This makes them more vulnerable to mistreatment and more likely to be displaced from their lands.

These logics took over land and turned people and nature into resources that would be harnessed towards one goal – the accumulation of capital. Historians, environmental studies, Indigenous and post-colonial scholars note that colonial logics were particularly successful because of their ability to alter ecosystems. As the examples below will attempt to demonstrate, colonists, in their rush to meet the consumer needs of their home countries, brought in foreign species that were invasive or completely overtook certain landmasses while driving out Indigenous people who could not defend themselves against biological invasion.[23] Eduardo Mondaca sees four important themes in the global pattern of power.[24] The first is the racialization of relations between colonizer and colonized such that the relationship of domination is naturalized. The second is the creation of an exploitative system that links the control of labour and land in order to produce goods for the market. The third is the locus of Europe as both the centre of production and knowledge. Finally, it is the establishment of a controlling system of collective authority, with control given to those seen as superior. This last is a further important point when aware of the collusion of 'local' elites in systems of domination and colonialism. As Martin Mahony and Georgina Endfield note, ideas of empire were intimately bound with ideas of climate – a relationship perhaps best epitomized, they say, by Montesquieu's observation that 'The empire of the climate is the first, the most powerful of all empires'.[25] These historical processes of extraction, dispossession, replacement and purposeful extinction drove colonization and ecological imperialism as part of the moral order of modern capitalism.[26] The idea of tropical abundance held that tropical areas, with their hot, humid climates and rich soils, were highly biologically productive, and that little human work was required to grow subsistence crops. Travellers like Sir Walter Raleigh sent home reports of tropical inhabitants eating only that which 'nature' brings forth, and luxuriating amid the redundancy of hard agricultural labour.[27] These ideas, solidified in the sixteenth century, had lasting effects on imperial policies and practices, perhaps most notably in taxation schemes which were designed to compensate for the over-supply of cheap, productive land which was thought to keep local wages too high. Such abundant lands were nonetheless attractive places to take control of and exploit. Commercial monocultures were spread through land-grabbing, wars and slavery.[28] As Jason Moore has described, the production and distribution of certain commodities, such as sugar, silver, gold mining, tobacco, coffee, cotton and grain were privileged, and the process of doing so meant that geographic spaces in the 'margins of the world-system' were restructured in order to require further expansion.[29] Many of the grains and animal products being exported from Australia, New Zealand, Argentina and Uruguay, for example, are products that were completely foreign to the landscape before colonization. Dal Lago, providing a comparison of American plantations and European landed estates and how they expanded capitalist space, also notes how American nineteenth-century agricultural production on large-scale plantations was also strongly related to the creation of slave 'commodities' [sic] in the Southern United States, Cuba, Brazil and much of the Atlantic.[30] Laws for labour and conservation were drawn up to codify such processes that emptied lands and displaced indigenous nature and communities, all the while mobilizing the accumulation of capital.[31] In 1884 and 1885, the Berlin Conference, also known as the West Africa Conference and organized by Chancellor Bismarck, resulted in the division of the African continent into colonies, with the different European states asserting their sovereignty over each one. This process was key to the legitimization of minerals from African countries. Most historical work referring to this period notes how European nations – especially Belgium, France, the United Kingdom and the Netherlands – imposed these laws on African colonies, justifying the violence meted out to people and nature as part of the civilizing mission of modernity.[32] It was quite usual for armed groups to be funded by extractive companies. Going back to Casement's narrative, we have the example of rubber extraction companies who created security forces from enslaved Indigenous youngsters, forcing them to discipline their enslaved brethren. In *Culture and Imperialism*, Edward Said rightly argues, therefore, that the purpose of empire was the possession of land.[33]

Gurminder Bhambra agrees, pointing to the accumulation of private property as the first stage of colonialism. She notes that

> the initial processes of colonization in the New World were undertaken by private companies that were given charters to explore, to seek profits, and to obtain lands while also making claims for jurisdiction and sovereignty in the name of various European monarchs as well as the Pope.[34]

This was followed, in the sixteenth to eighteenth centuries, by the elimination and dispossession of Indigenous lands.[35] Brenna Bhandar, building on this, agrees, noting that the insistence on private property within Western culture is linked to this desire for the possession of land and use of bureaucracy in order to disenfranchise local communities.[36] As such, Patrick Wolfe argues that the process was structural and not one simple spontaneous event.[37] International laws made it legal to appropriate and expropriate land and also people. As Anibal Quijano has discussed, there is a fundamental link between the coloniality of knowledge and economic coloniality.[38] These laws were based on epistemic social classifications, or racialization, such as making the African inferior within Christian cosmology in order to legitimatize the slave trade.[39]

If I am to discuss colonial logics in this chapter, I cannot escape the reality that Christianity played a key part in structuring the civilizing aspects of this mission, linked to identity markers prescribed by European Christianity and white masculinity.[40] Christian missionaries were important in preparing colonized lands for European settlement, as well as preparing colonized bodies for slavery. The Absolute Abrahamic God was suggested as the foundation of all meaning, a totalizing God to whom all other meanings and knowledge were made subservient. It suggested a social order that was dependent on the edification of the colonizer.[41] Sylvia Wynter brilliantly teaches us the importance of the concept of race in colonial regimes of social domination. As Wynter notes, these logics found a way for the Western actor to replace distinctions of mortal/immortal, natural/supernatural, gods/Gods as the actor on whose 'basis all human groups had ... grounded ... their descriptive statement of what it is to be human'.[42] The colonizer was more than human and the barbaric heathen was the one in need of taming and containment. For example, consider the understanding of *maa-baap* in India. If God, presented as the Absolute parent, was the accepted discourse, it was then easy to extend this to the metaphor of empire as a 'family'. Here the emperor or empress was the *maa-baap* (mother and father) to the untrained native child. The empire then is the drawing room where the native is trained.[43] Therefore, we have not only the legitimization of the disposability of land, but also the disposability of the labourer. Walter Mignolo, discussing this, notes that the genocide of Indigenous populations in the Americas was not only due to violent conquest or the diseases that the colonizers brought with them, but was also due to the fact that so many were forced to work until death.[44] Cheryl Harris further argues how whiteness was encoded as a value into land and social relations.[45] Looking at chattel slavery, Harris teaches us that the logic here was premised on the appropriation of Indigenous lands that allowed for Black slaves and other non-white persons to be seen as property. As such, the conception of race interacted and became intertwined with the conception of property and land in order to maintain material and spiritual domination. We are still within the grip of these dominations. Let me illustrate this with two examples: Indonesia and Sri Lanka.

Old sins and long shadows: plantations in Indonesia[46]

Let us take the example of plantations in both colonial and modern Indonesia. I borrowed the quotation in the title of this chapter from an article by the political geographer Lisa Tilley that examines European plantation ownership in Indonesia.[47] Along with the Congo, Indonesia was one of the two largest resource-rich tropical colonies for Europeans. In her work, what Tilley discusses, alongside her engagement with the rebellions of Indonesian plantation workers, is how industrial racial regimes were 'produced, distributed and fractured', resulting in an order of social relationships and environmental reality that are entirely exclusive. One of the significant interventions by the Dutch was the introduction of the plantation model.[48] The Dutch East India Company (VOC) used what was called the *cultuurstelsel* or enforced planting policy.[49] The Javanese peasantry were obliged to grow commercial crops like coffee, sugar cane, indigo, tobacco and tea, which were then handed over to the colonial administration.[50] In order to grow these crops, vast areas were deforested, as in Northern Sumatra. Studies contend that more than half the land in Deli Serdang and Langkat was cleared for plantations.[51] The Dutch used a system of forced cultivation that ensured that the local peasant production was inextricably tied to industrial manufacture. Small landholders often lost their lands through a systematic method of displacement by European plantation owners, who were often aided by complicit Javanese village chiefs.[52] Tilley, quoting Leo Waibel, notes the ways in which the Dutch refused to credit the Indigenous population with agricultural skills, due to their 'lack of scientific and technical knowledge'.[53] The plantation system was implemented as a way in which to bring together highly unskilled labour with highly technical knowledge.

A key example within Indonesia is also the Dutch pursuit of nutmeg in the Banda Islands and the establishment of a colonial order that had, as its main purpose, intensive silvaculture.[54] Amitav Ghosh's majestic work *The Nutmeg's Curse* describes the burning of a Bandanese village by officials of the Dutch East India Company as they pursued a monopoly over nutmeg extraction.[55] Ghosh remarks, as all the authors I invoke in this chapter do, that the Dutch East India Company was an exemplar of early capitalism in that they relentlessly understood the world as *resource* so that they were driven by a need for omnicide – a need to destroy everything.

In the Banda Islands, the Dutch established the *perkenier* system of slavery, which Antony Reid describes as unheard of in Asia – a New World style of slavery.[56] So that they did not have to be dependent on informal growers of nutmeg trees, the Dutch East India Company (VOC) put together a plan that would repopulate and reorganize the Banda Islands. The Dutch killed, removed or enslaved the Indigenous inhabitants of the islands and replaced them with the slaves.[57] The land was divided into plantations and called 'perken', meaning gardens. *Perkeniers* who managed the land were obliged to sell the nutmeg and the mace that they harvested for a price determined by the VOC. Each *perkenier* had 25 slaves allocated to them. This enslaved population came not only from other Molucca islands but also from Sri Lanka, India, Malaysia and the Cape Colony. Often prisoners from Java and Makassar were also added to this group.[58] Another example is Jacob Nienhuys, a Dutch tobacco trader who was looking to develop plantation land in Deli. Faced with a lack of labour, Nienhuys 'imported' thousands of Chinese 'coolies' from Malaysia, and Singapore.[59] Workers were also brought in from Java, Banjar and India. This system of indentured labour is, of course, a common characteristic of European colonial logics where labour from one country was sent to work in another. Struggling with the problem of labour shortages and a need to extract from fertile land, the 'coolie' system of 'unfree' labour sprang up. 'Coolie' refers to an indentured service that aims to provide cheap and well-controlled labour. Its use followed the dismantling of slavery, and the establishment of a modern racial governmentality of 'free' yet racialized and coerced labour.[60]

In the seventeenth and eighteenth centuries, European planters in the United States shopped European and Chinese labourers on their plantations. As Sunanda Sen notes, this became more prevalent in the 1800s once slavery became abolished in Britain and Europe. Large-scale shipments of Indian labourers, for example, were organized on short-term contracts by British commercial and financial agencies to work on British plantations in Africa and other parts of Asia.[61] Although these were short-term contracts with wages – though low paid – humanitarian activists of the mid-1800s decried this as a new form of slavery, especially noting what they felt was a kind of bonded labour.[62] This was on top of the eviscerations and effects on the Indigenous population and land in the area. Housing and feeding indentured labour also required the destruction of land and the driving out of the local population.[63]

This, then, is an example of how empire not only reorganized land but also reorganized society. I repeat, there is no way of separating the violence done to colonized and enslaved people from the violence done to the land. As Waibel concludes, it is the industrialization (of both people and land) that, even though unsuited to the agriculture of the 'temperate zones', is the most important characteristic of the plantation system.[64] What the VOC followed by the Dutch government implemented then is still replicated in the palm oil plantations in Indonesia today.[65] Indonesia supplies over half of the world's palm oil, with monocrop palm oil plantations that cover millions of hectares of land. Although primarily found on the African continent, palm oil plants thrived in Asia and were easy to manage. According to historians and geographers, by 1940 Indonesia and Malaysia were exporting more palm oil than Africa.[66] Palm oil is ubiquitous in our daily lives because it is cheap and it was made so by colonial policy that not only exported the plant to Asia but also brought in 'coolie' labour to work on the plantations. After independence, plantation companies retained their cheap and easy access to this land, resulting in the situation we have today. Human Rights Watch reports note that between 2001 and 2017, Indonesia lost 24 million hectares of forest cover to palm oil cultivation.[67] This is an area about the size of the United Kingdom. Tania Li and Pujo Semedi's impressive ethnographic work on these plantations makes the link between the plantation as an occupying corporate force and how this 'rooted' plantation is able to draw on colonial-racial logics that are firmly embedded in Indonesian law and political discourse.[68] They note how palm oil, being in demand much in the way that sugar was in earlier centuries, requires a multiple, extractive regime that creates profit for both capitalist conglomerates and the Indonesian state. In their expansive work, Li and Semedi underline how plantation workers are often manipulated into unfair contracts and trafficked across borders, a situation that is difficult to change due to grandfathered (exempted) labour laws.[69] In fact, they also point out that unfree labour was very common during the colonial period, and the current labour regimes are often worse because there is also the use of casual and contract labour but very little in the way of labour rights due to the short-term nature of these contracts. In the meantime, there is little income for those who work the plantations because they continue to be forced into a way of life beyond their control.[70] Paul Gellert agrees, noting how accumulation by dispossession, that is, the colonial practice of land-grabbing, is used by corporations in order to expand palm oil plantations.[71] Gellert further notes that accumulation by dispossession is now an accepted tactic of corporate neo-liberalization that is supported by a domestic state and local elites. As with all long histories of production and dispossession, it is not only the environment but also the Indigenous populations that are most affected by these ongoing logics. Reports collated by community activists for Amnesty International and Human Rights Watch note that there is little evidence that communities are consulted by multinational corporations when plantations are expanded.[72] Often communities only realize that their lands and forests are going to be destroyed when the bulldozers turn up.[73] There is little or no meaningful consultation with Indigenous people, who immediately lose not only their livelihoods and ways of living, but also experience the breaking down of their culture, languages, knowledge and unique traditions which are deeply embedded with the natural environment.

What's in your breakfast cup? The case of Sri Lanka

Tea is not a drink of solidarity and neither is coffee. In Kandy, a city in the Central Highlands of Sri Lanka, you might come across a mural telling you the story of a Scotsman and the planting of tea in the country. This obliterates the long history of peasant smallholders who grew coffee for many decades before the arrival of plantations. Coffee was grown in what you might call your home garden as part of a supplementary income.[74] The fact that coffee grew successfully in these areas encouraged European settlers in the 1820s to then establish large-scale coffee and tea plantations or what are called estates. Key to this was the creation of a planter-official class. Civil servants were allowed to make up for their loss in salaries and pensions through investments in commercial land. The Crown Lands Encroachment Ordinance, a rather notorious bill passed in the 1840s, allowed for the colonial state to claim any uncultivated or unoccupied land. Any local or Indigenous people wishing to lay claim to this land could only prove ownership according to the stringent criteria laid down by the state. These conditions were often impossible to meet. This land was then demarcated as Crown land and sold to planter civil servants by the acre.[75] In this way, by dispossessing the peasant and Indigenous communities, the British established in Sri Lanka a planter-official class restricted to civil servants, religious leaders and village headmen.[76] Gone were small-scale agricultural home gardens that interfered little with the natural ecology, and in were brought large-scale measures for land that could now produce cash crops to be sold to Europe. Nira Wickramesinghe's in-depth work on Sri Lankan history details how financial conditions were created in order to promote the new crop. Loans, agency houses and financial assistance were provided to planters, as well as well-defined rights of ownership.

Lal Jayawardena notes that by the 1870s the colonial administration had decided on a repressive attitude towards traditional cultivation.[77] The primary purpose of this was to stop the potential revenues being raised by village headmen and then to place on the market areas that coffee planters were looking for. As the demand for land grew, there was increasing deforestation, as well as an encroachment on *chena* (traditional slash and burn) cultivation. Historians note that the Crown used the 1840 Ordinance to claim all uncultivated or periodically (traditionally) cultivated land by demarcating it as waste land. Waste land was then cleared, tamed and reordered into productive land. Peasants and Indigenous communities were thwarted by a new discourse of 'property rights' and the requirement of documentary proof of such rights. As most land sales had previously been handled through oral agreements, the Crown was easily able to step in to seize lands it could then sell to planters. The moral reason given for this change was the government's wish to redirect the villager to a more civilized activity – that of paddy cultivation. This repressive policy did speed up land sales by impoverished villagers, who would sell their lands to village headmen; however, this often resulted in various disputes. As a result of these policies of containment and dispute, by the early 1890s large slices of the central belt had been turned into tea and rubber estates.[78]

In the first instance, this triggered large-scale erosion of nutritious soil. Studies by UNESCAP that looked at tea, rubber and coconut plantations in Sri Lanka found that areas where tea and coffee are planted are highly vulnerable to erosion.[79] This erosion stripped nutrients and topsoil from the agricultural fields and created issues downstream. UNESCAP further noted that siltation from erosion, which fills reservoirs and therefore reduces hydropower generation, is a significant problem in Sri Lanka. In the second instance, the influence of these containment and land sale policies are still found within Sri Lankan law today. A variety of studies on land rights, including one on which I was a research associate in the years 2017–18, noted that in recent years 17,273 hectares of land have been appropriated by the government for agricultural projects.[80] Most of these are for large-scale plantation projects for the production of bananas, cashews and corn for international export. Forest land has also been seized and cleared for these projects, leading to the loss of watersheds, elephant corridors and migration routes for wildlife. Deforestation has proved to be the most detrimental predicament for Sri Lanka because of the large number of forests that had been eliminated for more plantations and agricultural land. Sri Lanka is a big and major producer of tea, and the deforestation has provided enough room for these tea plantations to produce this significant resource and raise the country's revenue. Although the removal of Sri Lanka's forests has provided land, it has also caused many environmental problems. For example, deforestation has caused soil erosion, landslides and flooding, and has effectively damaged the wildlife and biodiversity in these forests. Furthermore, maintaining the remaining environmental forests has also proved to be a dilemma because of Sri Lanka's demand for and dependence on timber for fuel, and the fact that its government policies are heavily focused on tree plantations and timber production. Land is also easily appropriated for tourism projects, destroying coastal ecosystems and the livelihoods of farms and fishermen. It is the mobilizations of some of these groups that I discuss in the next chapter.

In the final instance, we once again come up against labour, and this is certainly the enduring legacy of the plantations. Labour was a key requirement for such large-scale production and the local communities, resentful of the loss of their land as well as the lack of any assurance of the protections they had received under the previous monarchical system, were reluctant to be employed by the planters. Wages were also very low. Nira Wikcremasinghe and Asoka Bandarage also note that very little effort was made to recruit local labour. As a result, the British then began bringing in foreign labour.[81] Thousands of Indian 'coolie' labourers from the south of India were brought in to the central and southern plantations. Few regulations governed the recruitment process because it was considered to be the movement of 'free' labour. In reality however, these workers were not afforded even the minimal protection provided for workers on indenture. Workers often fell ill – fever and dysentery prevailed on the estates – and workers were not sent to hospital or given treatment by medical advisors.[82]

These labourers were subject to low wages, long hours of work and substandard accommodation. The planters also brought in labourers from mixed caste backgrounds, using these caste hierarchies to manipulate labour relationships. As supervisors were brought in from higher castes and field labourers from lower ones, it ensured strict discipline. It also ensured that these groups would not team up with each other to destabilize the European planters. Unfortunately, even after independence this practice of using caste as a means of control extended into the new plantation regimes, severely affecting ethnic relations and political mobilizations.[83] In a 2021 UN Special Rapporteur statement, these labourers were noted as being part of a contemporary form of slavery. The Special Rapporteur noted that these communities were still facing forms of exploitation and abuse and were still living in the line houses built by planters over 200 years ago. Mythri Jegathesan's intensive *cami pakkiratu* narrative into the lives of these Up-Country or Malaiyaha Tamils noted that although they are often eclipsed within representations of the country's civil and political conflict, they have experienced protracted forms of discrimination that directly result from social and economic matrices of escalating civil violence, legal and affective exclusion, and neo-liberal policies of worker dispossession.[84] Unrecognized as citizens by both the Indian and Sri Lankan governments, this community was long denied identification documentation, a key piece of paperwork dictated by now enshrined leftover colonial law, which is required in the country should you wish to own property. Further to this, feminist scholars have noted that there exists a kind of plantation patriarchy that has its roots in the systems of control and discipline implemented during the colonial era.[85] Plantation patriarchy was directly linked to the profitability of production, and it was therefore in the interests of those in power to sustain norms and practices that promoted low status and therefore lesser entitlements for women workers. The power of 'plantation patriarchy' extended beyond the boundaries of the unit of production, and planters exerted their influence on the state and wider society, gaining favourable concessions and legal support. Women were part of the labour force recruited by the planters from the 1830s from the famine-prone districts of the Madras Presidency in South India. These women belonged to the lowest classes and castes in the villages and, like their male counterparts, suffered from chronic indebtedness and sought some form of remuneration for their survival. These workers were, as noted by Sir Emerson Tennent, the Colonial Secretary in charge of Ceylon in 1847, caught between the dictates of plantation capitalism and debt bondage, with little choice of place of work or seasonality of employment.[86] In more recent decades, we have seen union and political organizing by these communities, especially the women's federation, but there yet remains a long history of legal and social wrangling that requires transformation.

Conclusions

It is difficult to think how to conclude such a chapter. In some ways, perhaps the best thing is simply to leave this with its narration, highlighting the key case studies as vignettes of history and urging the reader to engage in their own exploration of the links between ecological injustice, labour exploitation and the crisis we find ourselves in. Perhaps a better focus for the chapter might have been to carry out the exercise I often use with undergraduate students, which is to ask them to investigate where the minerals in their mobile phones come from. What is the history? What is the present? At the same time, one also feels that what is provided here is, in many ways, history that is known, even though it is history that is difficult to acknowledge. Not only this, a key mobilizing point for ecological justice movements is to move against local and national elites who continue colonial policies that subjugate land and people. It is also to recognize and struggle against the global political economy that still cries out for mass-produced goods like tea, coffee and palm oil, thus legitimizing the rapacious activities of national and multinational entities. Is there a way to effect a moral awakening without drastically transforming our current ways of living? In recounting these brief histories and recycling the arguments of how the political economy was made, I am not attempting to elicit grief or guilt. Yes, we must reflect and acknowledge our sinfulness, but this recognition is about being motivated to action, to transformation, to the struggle to achieve the full humanity that was discussed in Chapter 1. A key ingredient in doing so is to be aware of the mobilizations and struggles of those who are in the thick of this battle. In the next chapter, I discuss these mobilizations, looking at ecological justice movements, especially faith-based ones, in Canada, the UK and Sri Lanka. This will suggest to us the kinds of rainbow coalitions at play here, the intersections of the struggle in the Global South, but also the difficult 'whiteness' that still plagues many of these movements. This last links to an epistemic issue also discussed in this chapter – the cultural urge to fetishize nature in such a way that it is denied agency, but is also placed in a state of grace over disposable bodies.

Notes

1 Lisa Tilley, '"A Strange Industrial Order:" Indonesia's Racialized Plantation Ecologies and Anticolonial Estate Worker Rebellions', *History of the Present* 10, no. 1 (April 2020), https://www.dukeupress.edu/history-of-the-present.

2 Amitav Ghosh, 2017, *The Great Derangement: Climate Change and the Unthinkable*, The Randy L. and Melvin R. Berlin Family Lectures, Chicago, IL: The University of Chicago Press.

3 The phrase 'civilized savagery' I borrow from Aime Césaire, who condemned in the strongest terms Europe's savage exploitation of the non-European world's labour and resources. Césaire famously termed colonization as 'thingification'. In *Discours sur le Colonialisme*, Césaire argues over and over again how colonization, and the concomitant acts of violence, rape and torture that go with it, degrades Europe itself. The colonizer becomes barbarian but is also dependent on turning the Other into a barbarian.

4 Frantz Fanon and Jean-Paul Sartre, 2001, *The Wretched of the Earth*, trans. Constance Farrington, Penguin Classics reprint, London, New York: Penguin Books.

5 Robin D. G. Kelley, 'A Poetics of Anticolonialism', *Monthly Review* (1 November 1999), https://monthlyreview.org/1999/11/01/a-poetics-of-anticolonialism/. Kelley translated the English version of *Discours*.

6 Aimé Césaire, 2008, *Discours sur le Colonialisme*, Nachdr, Paris: Présence Africaine.

7 In this way, it is a kind of vice versa of James Cone's argument, presented in the previous chapter, that the struggle for racial justice must also include the struggle for ecological justice.

8 I am always grateful to the work of Kumari Jayawardene in differently educating me – an excellent example is Kumari Jayawardena and Rachel Kurian, 2015, *Class, Patriarchy and Ethnicity on Sri Lankan Plantations: Two Centuries of Power and Protest*, Critical Thinking in South Asian History, Hyderabad, Telangana, India: Orient BlackSwan. More recently, Mythri Jegathesan's 2019 work *Tea and Solidarity* (University of Washington Press) is an ethnographic tour de force in documenting plantation life and the contemporary struggles of plantation workers in Sri Lanka to reclaim their identity.

9 See, for example, Nindyo Sasongko, 'Epistemic Ignorance and the Indonesian Killings of 1965–1966: Righting the Wrongs of the Past and the Role of Faith Community', *Political Theology* 20, no. 3 (3 April 2019), pp. 280–95, https://doi.org/10.1080/1462317X.2019.1568705, in which Sasongko details how epistemic ignorance denies the work towards justice for the victims of Indonesian killings. They see an important role for faith communities in memorializing these events,

lamenting them and therefore making known the complex, hard histories of modern Indonesia.

10 James Baldwin, 2017, *The Fire Next Time*, Penguin Modern Classics, London: Penguin Books.

11 Gurminder K. Bhambra, 2014, *Connected Sociologies*, Theory for a Global Age, London: Bloomsbury.

12 Michel-Rolph Trouillot and Hazel V. Carby, 2015, *Silencing the Past: Power and the Production of History*, Boston, MA: Beacon Press. See also Harini Amarasuriya et al., 2020, *The Intimate Life of Dissent: Anthropological Perspectives*, London: UCL Press.

13 James Miles, 'Historical Silences and the Enduring Power of Counter Storytelling', *Curriculum Inquiry* 49, no. 3 (27 May 2019), pp. 253–59, https://doi.org/10.1080/03626784.2019.1633735.

14 Edward W. Said, 1995, *Orientalism: Western Conceptions of the Orient*, reprinted with a new afterword, Penguin History, London: Penguin Books.

15 Leonardo Boff, 1995, *Ecology & Liberation: A New Paradigm*, Ecology and Justice Series, Maryknoll, NY: Orbis Books, p. 43.

16 Ian James Kidd, José Medina and Gaile Pohlhaus (eds), 2019, *The Routledge Handbook of Epistemic Injustice*, Routledge Handbooks in Philosophy, London New York: Routledge.

17 Gustavo Gutiérrez, 1990, *The Truth Shall Make You Free: Confrontations*, Maryknoll, NY: Orbis Books.

18 Jose Madappattu, 2014, *Evangelization in a Marginalizing World with Special Reference to the Marginalised Satnamis in the Diocese of Raipur*, PhD dissertation, https://nbn-resolving.org/urn:nbn:de:101:1-2016110011845, accessed 14.06.2022; Sergio Torres and Virginia Fabella (eds), 1978, *The Emergent Gospel: Theology from the Developing World: Papers from the Ecumenical Dialogue of Third World Theologians, Dar Es Salaam, August 5–12, 1976*, London: Geoffrey Chapman.

19 Robbie Shilliam, 2015, *The Black Pacific: Anti-Colonial Struggles and Oceanic Connections*, Theory for a Global Age, London: Bloomsbury Academic.

20 Sankaran Krishna, 'Colonial Legacies and Contemporary Destitution: Law, Race, and Human Security', *Alternatives: Global, Local, Political* 40, no. 2 (2015), pp. 85–101, http://www.jstor.org/stable/24569425.

21 Augusto Javier Gomez Lopez et al., 2014, *Putumayo: la vorágine de las caucherías: memoria y testimonio*, Bogotá Centro Nacional de Memoria Historica.

22 Ann-Marie Agyeman and Anupama Ranawana, 2021, 'Exploring the Relationship between Race, Ethnicity, Colour and Poverty in Christian Aid's Work: Insights for Practice', London: Christian Aid.

23 Alfred W. Crosby, 2004, *Ecological Imperialism: The Biological Expansion of Europe, 900–1900*, Studies in Environment and History, 2nd edn, Cambridge: Cambridge University Press.

24 Eduardo Mondaca, 2017, 'The Archipelago of Chiloé and the Uncertain Contours of Its Future: Coloniality, New Extractivism and Political-Social Re-Vindication of Existence', in David Rodríguez Goyes et al. (eds), *Environmental Crime in Latin America: The Theft of Nature and the Poisoning of the Land*, London: Palgrave Macmillan UK, pp. 31–55.

25 Martin Mahony and Georgina Endfield, 'Climate and Colonialism', *WIREs Climate Change* 9, no. 2 (March 2018), https://doi.org/10.1002/wcc.510.

26 Farhana Sultana, 'Critical Climate Justice', *The Geographical Journal* 188, no. 1 (March 2022), pp. 118–24, https://doi.org/10.1111/geoj.12417.

27 Horatio John Suckling, 2018, *Ceylon: A General Description of the Island, Historical, Physical, Statistical*, New Delhi: Asian Educational Services (originally published 1876), p. 19.

28 Su-ming Khoo, *Colonial Dispossession and Extraction*, 'The Making of the Modern World', Lecture Series (Connected Sociologies, 9 November 2020).

29 Jason W. Moore, 'Madeira, Sugar, and the Conquest of Nature in the "First" Sixteenth Century: Part I: From "Island of Timber" to Sugar Revolution, 1420–1506', *Review (Fernand Braudel Center)* 32, no. 4 (2009), pp. 345–90, http://www.jstor.org/stable/41427474.

30 Enrico Dal Lago, 'Second Slavery, Second Serfdom, and Beyond: The Atlantic Plantation System and the Eastern and Southern European Landed Estate System in Comparative Perspective, 1800–60', *Review (Fernand Braudel Center)* 32, no. 4 (2009), pp. 391–420, http://www.jstor.org/stable/41427475.

31 Su-ming Khoo, *Colonial Dispossession and Extraction*.

32 Gustavo Rojas-Páez, 2017, 'Understanding Environmental Harm and Justice Claims in the Global South: Crimes of the Powerful and Peoples' Resistance', in David Rodríguez Goyes et al. (eds), *Environmental Crime in Latin America: The Theft of Nature and the Poisoning of the Land*, London: Palgrave Macmillan UK, pp. 57–83.

33 Edward W. Said, 1994. *Culture and Imperialism*, London: Vintage, p. 35.

34 Gurminder K. Bhambra, 'Colonial Global Economy: Towards a Theoretical Reorientation of Political Economy', *Review of International Political Economy* 28, no. 2 (4 March 2021), pp. 307–22, https://doi.org/10.1080/09692290.2020.1830831.

35 Robert Nichols, 'Theft Is Property! The Recursive Logic of Dis-

possession', *Political Theory* 46, no. 1 (February 2018), pp. 3–28, https://doi.org/10.1177/0090591717701709.

36 Brenna Bhandar, 2018, *Colonial Lives of Property: Law, Land, and Racial Regimes of Ownership*, Global and Insurgent Legalities, Durham, NC: Duke University Press.

37 Patrick Wolfe, 'Settler Colonialism and the Elimination of the Native', *Journal of Genocide Research* 8, no. 4 (1 December 2006), pp. 387–409, https://doi.org/10.1080/14623520601056240.

38 Aníbal Quijano, 'Coloniality and Modernity/Rationality', *Cultural Studies* 21, no. 2–3 (March 2007), pp. 168–78, https://doi.org/10.1080/09502380601164353.

39 J. Kameron Carter, 2008, *Race: A Theological Account*, Oxford; New York: Oxford University Press.

40 Interestingly, as Ben Phillips notes, Christian poets like Milton and Donne both provide lyrical justification for why the land should be dominated. Female poets like Aphra Behn and Margaret Cavendish provide ripostes to Donne and Milton, noting the awareness within seventeenth-century culture of the downgrading of land, and subsequently the downgrading of women. In Phillips' work on ecofeminist theory, this awareness is part of a long culture of women identifying themselves with the earth's sufferings. Of course, these women poets do not also identify themselves with communities of colour who were treated in the same way as disposable land. As Kochurani Abraham has argued, Edwardian and Victorian feminists still required the pitiable native – especially the native woman – to be a figure upon whom they could build their feminist activism. Hence, they are not to be seen as equal with white women. This is an interesting occurrence considering also that much of the critique of environmental justice movements and even eco-theology – as we shall see in Chapter 4 – also points to these movements' fetishization of nature, of the Anthropocene.

41 Kirsten T. Edwards, 'Christianity as Anti-Colonial Resistance?', *Souls* 15, no. 1–2 (1 January 2013), pp. 146–62.

42 Sylvia Wynter, 'Unsettling the coloniality of being/power/truth/freedom: Towards the human, after Man, its overrepresentation – An argument', *CR: The New Centennial Review* 3, no. 3 (2003), pp. 257–337.

43 Zohreh T. Sullivan, 1993, *Narratives of Empire: The Fictions of Rudyard Kipling*, Cambridge: Cambridge University Press.

44 Walter Mignolo and Catherine E. Walsh, 2018, *On Decoloniality: Concepts, Analytics, Praxis*, Durham, NC: Duke University Press; Katharine Gerbner, 2018, *Christian Slavery: Conversion and Race in the Protestant Atlantic World*, Philadelphia, PA: University of Pennsylvania Press, https://www.degruyter.com/doi/book/10.9783/9780812294903.

45 Cheryl I. Harris, 'Whiteness as Property', *Harvard Law Review* 106, no. 8 (1993), pp. 1707–91. https://doi.org/10.2307/1341787.

46 In their excellent study *Plantation Life*, Li and Semedi define plantations as a machine for assembling land, labour and capital under centralized management for the purpose of making a profit; it is also a political technology that orders territories and populations, produces new subjects and makes new worlds. I choose to adhere to this definition.

47 See Tilley, 'A Strange Industrial Order'. The Dutch – via the Dutch East India company – were present in Indonesia from somewhere around the 1600s. Van Der Kroef notes that the Dutch East India company had their own administrative and judicial machinery, currency and flag and were powerful enough to hold life and death over both their officials and those that they colonized. At the end of the eighteenth century the company was in severe decline due to the financial as well as political costs of maintaining a colonial as well as commercial enterprise simultaneously. In 1796, after the abolition of the Dutch East India company, the Dutch government gained full control over Indonesia. This occurred through a process of imperial revision, whereby the company and the territories it had taken over (beyond Indonesia as well) were absorbed by governments. This period marks a time in Europe overall when the colonies stopped being solely the playground of trading companies and became the national territory of European governments.

48 Ulbe Bosma, 2013, *Sugar Plantation in India and Indonesia: Industrial Production, 1770–2010*, Studies in Comparative World History, Cambridge: Cambridge University Press.

49 Once the Dutch government took over in 1796, they changed this system for a 'liberal' economic system, mostly to open up opportunities to private companies. This, however, created other systems of enforced labour.

50 C. Fasseur and R. E. Elson, 1992, *The Politics of Colonial Exploitation: Java, the Dutch, and the Cultivation System*, Studies on Southeast Asia, Ithaca, NY: Southeast Asia Program, Cornell University.

51 Karl J. Pelzer, 1978, 'The Problems of Converting Agricultural Concessions into Long Leases', in *Planter and Peasant: Colonial Policy and the Agrarian Struggle in East Sumatra 1863–1947*, Vol. 84, Leiden: Brill, pp. 86–121.

52 Interestingly, in their work *History of Christianity in Indonesia*, Karel Steenbrink and Jan Aritonang note that 'worker missionaries' in Indonesia became critical of the mother country when they observed this process of land-grabbing. Such worker or grassroots missionaries chose to side with the people against oppression by witnessing to this violence. What solidified this also, argue Steenbrink and Aritonang, was

the fact that many of these missionaries originated from lower classes in their own country who had not yet been given the right to vote. The superiors who drove the colonial system came from the upper middle classes and therefore there was a kind of class solidarity that developed between small landholders and grassroots missions. Unfortunately, any such criticism was often heavily censored in mission magazines. There is probably a need to study further how such alliances were censored throughout Christian history.

53 Leo Waibel, 'The Tropical Plantation System', *The Scientific Monthly* 52, no. 2 (1941), pp. 156–60, http://www.jstor.org/stable/17375.

54 David R. Carlson and Amy Jordan, 'Visibility and Power: Preliminary Analysis of Social Control on a Bandanese Plantation Compound, Eastern Indonesia', *Asian Perspectives* 52, no. 2 (2013), pp. 213–43, http://www.jstor.org/stable/24569821.

55 A. Ghosh, 2021, *The Nutmeg's Curse: Parables for a planet in crisis*, London: John Murray.

56 Anthony Reid, 2015, *A History of Southeast Asia*, Chichester: Wiley Blackwell.

57 L. H. Roper, 2017, *Advancing Empire: English Interests and Overseas Expansion, 1613–1688*, New York: Cambridge University Press.

58 Phillip Winn, 'Slavery and Cultural Creativity in the Banda Islands', *Journal of Southeast Asian Studies* 41, no. 3 (2010), pp. 365–89. https://doi.org/10.1017/S0022463410000238.

59 Budiman Minasny, 2020, 'The Dark History of Slavery and Racism in Indonesia during the Dutch Colonial Period', *The Conversation*, 2 July, https://theconversation.com/the-dark-history-of-slavery-and-racism-in-indonesia-during-the-dutch-colonial-period-141457#:~:text=Slave%20trading%20was%20widely%20carried, accessed 19.02.2022.

60 Lisa Lowe, 2015, *The Intimacies of Four Continents*, Durham, NC: Duke University Press.

61 Sunanda Sen, 'Indentured Labour from India in the Age of Empire', *Social Scientist* 44, no. 1/2 (2016), pp. 35–74. http://www.jstor.org/stable/24890231.

62 Ashutosh Kumar, 2017, *Coolies of the Empire: Indentured Indians in the Sugar Colonies, 1830–1920*, Cambridge: Cambridge University Press.

63 Jonathan Robins, 'Shallow roots: the early oil palm industry in Southeast Asia, 1848–1940', *Journal of Southeast Asian Studies* 51, no. 4 (2020), pp. 538–60, doi:10.1017/S0022463420000697.

64 It should be noted that Waibel's work does not critique the plantation model but describes it.

65 Hariati Sinaga, 2021, '*Buruh Siluman*: The Making and Maintaining of Cheap and Disciplined Labour on Oil Palm Plantations in Indonesia', in Maria Backhouse et al. (eds), *Bioeconomy and Global Inequalities: Socio-Ecological Perspectives on Biomass Sourcing and Production*, Cham: Springer International Publishing, pp. 175–93, https://doi.org/10.1007/978-3-030-68944-5_9.

66 Jonathan Robins, 2021, *Oil Palm: A Global History* (Flows, Migrations, and Exchanges), Chapel Hill, NC: University of North Carolina Press.

67 Human Rights Watch, 2021, 'Indonesia: Expanding Palm Oil Operations Bring Harm Government Fails to Protect Affected Communities, Environment', *Human Rights Watch*, 3 June, https://www.hrw.org/news/2021/06/03/indonesia-expanding-palm-oil-operations-bring-harm, accessed 14.06.2022.

68 Tania Li and Pujo Semedi, 2021, *Plantation Life: Corporate Occupation in Indonesia's Oil Palm Zone*, Durham, NC: Duke University Press.

69 Li and Semedi, *Plantation Life*.

70 Li and Semedi, *Plantation Life*.

71 Paul K. Gellert, 2015, 'Palm Oil Expansion in Indonesia: Land Grabbing as Accumulation by Dispossession', in *States and Citizens: Accommodation, Facilitation and Resistance to Globalization*, vol. 34, Current Perspectives in Social Theory, Emerald Group Publishing Limited, pp. 65–99, https://doi.org/10.1108/S0278-120420150000034004.

72 John Vidal, 2013, '"Indonesia Is Seeing a New Corporate Colonialism"', *Guardian*, 25 May, https://www.theguardian.com/world/2013/may/25/indonesia-new-corporate-colonialism, accessed 14.06.2022.

73 Human Rights Watch, 2019, 'Indonesia: Indigenous Peoples Losing Their Forests: Lack of Government Oversight, Corporate Accountability Affects Culture, Livelihoods', *Human Rights Watch*, 22 September, https://www.hrw.org/news/2019/09/22/indonesia-indigenous-peoples-losing-their-forests, accessed 14.06.2022.

74 Nira Wickramasinghe, 2014, *Sri Lanka in the Modern Age: A History*, updated 2nd edn, New York: Oxford University Press.

75 In Sri Lanka the struggle for land rights still goes on, and much of this is related to what is referred to as 'Crown Land'.

76 Asoka Bandarage, 'The Establishment and Consolidation of the Plantation Economy in Sri Lanka', *Bulletin of Concerned Asian Scholars* 14, no. 3 (1 September 1982), pp. 2–22, https://doi.org/10.1080/14672715.1982.10412654.

77 Ajit Singh, '*Lal Jayawardena (1934 – 2004) – a tribute on the first anniversary of his death*', Development and Change 36, no. 6 (1 November 2005), pp. 1219–23.

78 Eric Meyer, 'From Landgrabbing to Landhunger: High Land Appropriation in the Plantation Areas of Sri Lanka during the British Period', *Modern Asian Studies* 26, no. 2 (1992), pp. 321–61, http://www.jstor.org/stable/312678.

79 United Nations Economic and Social Commission for Asia and the Pacific, 2021, *Geospatial Practices for Sustainable Development in Asia and the Pacific 2020: A Compendium*, S.l.: United Nations.

80 Bin Yang and Jun He, 'Global Land Grabbing: A Critical Review of Case Studies across the World', *Land* 10, no. 3 (2021), p. 324, https://doi.org/10.3390/land10030324.

81 Asoka Bandarage, 2020, *Colonialism in Sri Lanka: The Political Economy of the Kandyan Highlands, 1833–1886*, Colombo: Vimukthi Press.

82 Patrick Peebles, 2001, *The Plantation Tamils of Ceylon*, New Historical Perspectives on Migration, London; New York: Leicester University Press.

83 Aruna Jayathilaka, 'Discriminations Created by the Structural Violence' (Case Study of the Tea Plantation Sector of Sri Lanka), 3 October 2014. Available at: https://ssrn.com/abstract=2504897, accessed 14.06.2022.

84 Mythri Jegathesan, 'Deficient Realities: Expertise and Uncertainty among Tea Plantation Workers in Sri Lanka', *Dialectical Anthropology* 39, no. 3 (2015), pp. 255–72, http://www.jstor.org/stable/43895153.

85 Maurits S. Hassankhan, Lomarsh Roopnarine and Radica Mahase, 2017, *Social and Cultural Dimensions of Indian Indentured Labour and Its Diaspora: Past Present and Future*, London: Routledge, available from https://www.taylorfrancis.com/books/9781315271712.

86 Rachel Kurian and Kumari Jayawardena, 2013, 'Plantation Patriarchy and Structural Violence: Women Workers in Sri Lanka', Legacy of Slavery and Indentured Labour Conference on Bonded Labour, Migration, Diaspora and Identity Formation in Historical and Contemporary Context, 6–10 June 2013, Paramaribo, Suriname.

4

A Rainbow Coalition: Faith-based Mobilizations in Northern and Southern Spaces

It is impossible to be in right relationship with the Divine if we do not, at the same time, work to be in right relationship with what the Divine has created and with the active and ongoing business of Creation. As the poet says, we live in a world that is stamped with the grandeur of the Transcendent One.[1] I have written in previous chapters about staring out at the ocean and the sadness of seeing plastic skirts of pollution. Staring out at that same ocean, watching deep turquoise waves pirouette and leap over rocks and fill pools, I feel a deep, frightening thrill. One senses both the beauty and peace of Creation, but also its destructive power. As the Orthodox priest Kaleeg Hainsworth notes, nature is both that which is resplendent with the Creator's beauty, but also still part of that which is fallen, and we must see the ongoing activity of redemption even in the soil.[2] This is a feeling I can only describe as an experience of grace. It is an encounter that does inspire the need to protect and care for Creation. Creation groans in its constant activity. There is, then, a worshipfulness, a liturgical sensibility even in environmentalist activism. What is difficult to navigate within this is the urge to fetishize nature without seeing the aspect of activity within Creation itself, as well as realizing the interwovenness of human and environmental liberation. All is fallen, so the work towards justice and liberation must occur in complex ways.

In this chapter I highlight the various, and sometimes complex, conversations and encounters I have had with activist, indigenous and environmentalist groups from North America, the United Kingdom and Sri Lanka. Some of these conversations occurred during fieldwork for a project of Collaborative Inquiry into the reception of *Laudato Si'* in North America.[3] When putting together some of the discussions within this chapter, I have used, with permission, the 'informal' encounters that happened after the formal data gathering. The rest of the conversations and encounters within this chapter are reflections from my own active participation in environmentalist activism, conversations that occurred while doing fieldwork for other pieces of research, or accompanying protests and demonstrations. When I began the process of writing this book, I also engaged several activists and faith leaders in structured conversations regarding how they discussed environmental justice issues within their faith communities. With respect to an ethics of care towards these conversations, and at their request, I have kept many of these encounters anonymous. I chose not to use a technological recording tool; instead I took copious notes and kept a reflection diary. In some of these conversations we prayed or reflected on Bible passages together rather than engaging in a structured back and forth. This method helped to supplement the reflections that I had collated through the last five years of engagement with faith-based environmental activism. What I wish to emphasize here is the complex array of environmental activism, particularly that which is faith based, arguing especially for the need to listen to what Indigenous and Global South groups are saying and how they choose to organize.

There is a significant element within what is expressed by these communities that is reminiscent of Ambalavaner Sivanandan's analysis of the Black Panther's revolutionary strategy.[4] Mainly this is the understanding that 'oppression everywhere is woven from the same fabric' and that there is a need to build a coalition of radical kinship – that is, of different voices, minorities and revolutionaries – in order to mobilize for complex liberations.[5] Importantly, many of these conversations were around the complicated question of what justice means. Justice will be different in each context and within each community. Justice, especially ecological justice, is something that is in flux and constantly negotiated.[6] To achieve this space of true justice, we must all, as Frantz Fanon would say, take on a process of *unsettling*.[7] What many of the conversations I had pointed out was the lack of such a differentiated justice narrative in the larger global justice movement regarding the climate crisis, a narrative that does not fully engage with the 'rainbow coalition' and the intersected mobilizations that are very present in environmentalism. It is in these voices that I see what I refer to as prophetic rage and they provide the key building blocks for the final chapter.

Environmental activism and faith: a brief overview

Environmental activism enjoys a long history, finding, in the Western world, popular expression as early as the 1960s.[8] This has occurred in different ways, with varying agendas and for a range of purposes.[9] Mobilizations to protect the natural world, as well as the ways of living of Indigenous peoples, have been present for far longer than that, as they are part of the work of resisting colonization and the appropriation of land.[10] Although I will discuss this in more detail below, it is important to underline, as the Indigenous scholar Eve Tuck does, that there is nothing more central for us to understand, especially in efforts to 'decolonize', that

> [Justice] involve[s] the repatriation of land simultaneous to the recognition of how land and relations to land have always already been differently understood and enacted; that is, all of the land, and not just symbolically.[11]

Within this chapter I wish to focus on faith-based activism, so I will not elucidate much further on more secular movements and networks, many of which have been studied in depth by other scholars.¹² A long history of activity is also true of faith-based activism regarding the environment, and I will provide a brief overview below.¹³ Importantly, in more recent times we have seen that environmental activism increasingly involves many different actors, linked through broad alliances and ranging from the local to the global.¹⁴ Religious entities pursue social justice goals across a broad range of issues, one of which is the issue of climate justice. Religious mobilizations regarding the environment are quite varied, from work to protect communities from environmental harm to ethical and theological reflection. We see religious people active in local protest movements, in transnational networks, and also advocating at the supra-institutional level such as the United Nations and the World Social Forum. Faith communities, especially those in the Global South, cannot shy away from the struggle for ecological justice because they live in areas that are already severely impacted by climate change. As Willie Jenkins notes, there are often tensions between religious actors in the Global South and their counterparts in, for example, the North Atlantic.¹⁵ From my own experience, having worked across the Christian INGO spectrum with CAFOD, Caritas Internationalis and Christian Aid, the pressing urgency to work for ecological justice comes from the twin impetuses of responding to the charges of the gospel as well as to the urgent demands of the communities that these agencies partner with for development projects. This need to work for a sustainable change is intensified by the fact that religious actors and faith-based agencies are also often rooted in their community, and are therefore not only a trusted local actor, but also have the necessary links to bring together a multiplicity of agents to advocate for policy change, or call attention to an issue on the world stage.¹⁶ The Joint Learning Initiative on Faith and Local Communities (JLIFLC) notes that faith-based NGOs, in particular, are active in working to mitigate climate change and mobilizing their communities towards advocacy for change.¹⁷ A study of 91 countries conducted during the years 1989–2014 noted that religion is often associated with a greater willingness to donate to or demonstrate for the environment.[18] This is more pronounced, it found, in lower-income countries.

As highlighted in Chapter 1, we can see statements from institutional religious bodies from as early as the 1980s. The example that I employed was the statement of the Filipino bishops, and in its essence it mirrors the work of the World Council of Churches which had a Climate Change Program in place before the Intergovernmental Panel on Climate Change was instituted in 1988. We continue to see strong statements from different Christian denominations, from national bishops' conferences within the Catholic Churches to mainline Protestant groups as well. As Robin Veldman et al. note, even conservative Christian groups such as the Seventh Day Adventist Church and the Southern Baptist Church have also issued statements or acknowledged the need for initiatives on this issue in recent years.[19]

This is not exclusive to Christian denominations. Let us consider some other examples. The Dalai Lama's first speech on climate change was in 1992.[20] Engaged Buddhists, working very much at the level of grassroots activism, often frame their mobilizations by speaking out against the religion of consumerism, using the *Addittapariyaya Sutta* (The Fire Sermon) as a key text.[21] For Buddhist environmentalists, there is a need to point out the indifference that is inculcated in consumerist culture that they then feel leads to the decimation of habitats, water pollution and, importantly, a failure to understand the deep interconnectedness of life.[22] The concept of stewardship and that human beings must be the *khalifah* (guardians) of the planet is a key framing used in the Islamic Declaration on Climate Change to propel change in environmental policy in Muslim countries.[23] Studies of environmental efforts within the Muslim world also note how, especially in East Asian countries, official Muslim institutions are placing pressure on governments to strengthen and justify environmental laws. There is also significant theological scholarship by Muslim academics noting that the empirical experience of the Abrahamic God is not one that is supremely transcendent or distant, and the need to bring theologizing together with religiously motivated activism.[24] The Rabbinic letter, written in 2015 by seven leading rabbis, argues, among other things, about the importance of realizing the links between worsening concentrations of wealth and power and the worsening climate crisis.[25] David Krantz's ethnographic work has noted how Jewish environmentalists are looking to change the observance of *shmita* by reclaiming aspects of environmental and social justice.[26] An exhibition of folklife at the Smithsonian Museum captured how a rabbi in the US used synagogue ritual and shared meals infused with environmental themes to incorporate ten environmental plagues into its Tu B'Shvat seder.[27]

Religious engagement is not only restricted to that which is formally recognized as *religion* per se. This can also refer to Indigenous communities that live within a spiritualist setting, and for whom, as mentioned before, the struggle for the preservation of the natural world, as well as ancient ways of living, has been a constant since the first moment that their lands were taken by an imperial force. Zoe Todd puts it pithily when she notes that not all humans are equally involved in the causes of the current climate crisis, and neither are these humans invited into the conceptual spaces where responses are formulated.[28] Todd quotes Andreas Malm and Alf Hornborg when they note that 'a clique of white British men literally pointed steam-power as a weapon – on sea and land, boats and rails – against the best part of humankind, from the Niger delta to the Yangzi delta, the Levant to Latin America'.[29] I have briefly discussed Indigenous theology in Chapter 2, but I wish to reiterate that such Indigenous mobilization is also a form of religious engagement, whether theological or not. As others have also argued, this is still religious engagement because there is here an understanding that the beyond-human world has cosmological and theological framings. Jenkins notes, for example, the intersection of ritual and ecological regulation in the Pacific Island cultures, as well as the Christian churches of Oceania and the importance of the story of Noah's ark to

Pacific Island Christianities. Within this understanding of cosmology is also the understanding of how natural bodies are worshipped as divinities, such as the river Ganga in India. I agree here with Jenkins that Indigenous communities often challenge the categories of religion, as well as the relationship between religion and climate change, especially with relation to culture. The following passage from Shekhar Pathak, which describes the cosmology of the Chipko community, is a beautiful example of this:

> From the knowledge passed on by ancestors through millennia and their lived experience, they have a palpable sensibility of mountains as animate beings imbued with enormous energy and complex emotions. They know the bountiful generosity and dangers which ensue when the Indian Ocean rises each year to embrace the voluptuous contours of the Himalayas – of the monsoons that erupt from this grand planetary coupling to swell the rivers and streams which renew life as well as trigger landslides and floods that destroy life. They know first-hand of the damage when the mountains quake and break apart as they rise towards the sky. They know the different kinds of trees and vegetation which protect the mountain slopes and valleys and enable the diversity of life to thrive and flourish.[30]

Vandana Shiva also articulates the cosmology that influences such movements as the Chipko movement, where there is an understanding of how the trees teach us unconditional love and unconditional giving.[31] This nature of the forest is the *aranya samskriti* (culture of the forest), which is a path of conscious choice. This is not an exclusive list, but one chosen to highlight the range of religious engagement.[32]

In most of this engagement, as referenced above, we see particular narratives that often appear. Chief among these is the need to protect and care for the natural world and being in right relationship with the universe. In the case of the Abrahamic religions, this is most often articulated in terms of the practice of worshipping the Creator, as well as protecting that which has been revealed to us. Most institutional religious actors aim to motivate concern among the faithful, while others, such as the Buddhist environmentalists noted above or a document like *Laudato Si'*, are also keen to criticize consumerist culture as well as the technocratic paradigm as a solution to the climate crisis. Another strong theme in faith-based activism is apocalyptic narrative. In the next section I will discuss briefly the difficulty caused by apocalyptic narratives and how they can – perhaps inadvertently – lend themselves to an imperialist environmentalism.

Apocalypse now ... in a minute?

A recent problem with environmental activism has been the turn to what we can call a kind of 'imperialist environmentalism'. In August of 2019, I co-wrote a short article for the sociology magazine *Discover Society* with the political theorist James Trafford. In this article we discussed the problem of 'imperialist environmentalism' and the varying struggles and discourses within environmental justice movements, including those that were within, or were led by, faith-based communities. The article also raised the problem of environmental justice movements and policies that seem indifferent to the long arm of history, and the need for reparative justice.[33] A key argument, stemming from my own ongoing empirical research at the time as well as Trafford's work on Britain's colonial legacies, was the danger for ecological justice movements like Extinction Rebellion and the school strikes, or for the various declarations of climate emergencies by countries in the Global North, to be hampered by a kind of 'white' myopia.[34] By this we meant an indifference to historical extraction and exploitation as well as a reluctance to foreground the knowledge and struggles of 'Southern' as well as Indigenous movements. In the previous chapter, I hope I have discussed in some depth the importance of understanding and being knowledgeable about imperial histories of extraction, violence and indebtedness.

Another discussion point in the article written with Trafford was how environmental struggles, especially those based in the Global North, can become part of a 'green nationalism' culture if they choose to retain such myopia. Our example was the spectacular mobilizations of Extinction Rebellion (XR) – of which a Christian XR exists – and their declaration that they sought to protect 'this nation, its peoples, our ecosystems and the future of generations to come'. It is discourse that looks at an eventual horizon of the apocalypse, and that is couched almost completely in terms of security.[35] The difficulty with this is, of course, the focus on an eschatological point. The effects of the climate crises may be a problem of the future for those of us who live in the Global North but it dismisses the very real experiences of those who are already baking, drowning and starving every day. One of our main concerns was also that there are movements like XR that see the climate crises and mass migration as intertwined dangers, with their former spokesperson Rupert Read going so far as to warn of events of large-scale immigration due to climate catastrophe.[36] That is to say that they urged governments to take the climate crises seriously lest the degradation of land in the Global South led to an influx of climate refugees. XR's green nationalism prompted a response from the coalition group Wretched of the Earth. I have already referenced this letter earlier and another excerpt provides a sharp rebuttal to the 'whiteness' of movements like XR:[37]

> You may not realize that when you focus on the science you often look past the fire and us – you look past our histories of struggle, dignity, victory and resilience. And you look past the vast intergenerational knowledge of unity with nature that our peoples have.[38]

It is not only XR that must be referred to here, but also the ease with which the far right is able to capture ecological discourse for itself, by seeing the climate crises as creating a scarcity of resources. I am discussing the case of XR to show how a group that is not on the far right per se can easily be appropriated by, or themselves use, oppressive ideologies. Nationalism is a project whose main obsession is the limiting of borders, and the climate crisis should be beyond borders. However, environmentalist discourses that argue in terms of food and resource security underline the importance of national borders and security. These are easily appropriated or aligned to far right agendas. As an example, we have seen in recent years how the previously climate denial trend of Marine Le Pen's far-right party is now picking up the idea of patriotic ecology.[39] For such groups, it is easy to make the connection between food and environmental security and the belief that different ethnic groups should remain distinct and restricted to their 'native lands'. Islamophobic groups in Europe, as well as South Asia, use animal rights discourse to morally chastise Muslim communities over their practice of halal.[40] In their 2021 report on the rise of hate speech in the UK, for example, the 'Hope Not Hate' group noted an increasing trend of green nationalism among right wing groups and newspapers. They note that,

> Particular strains of green thought have deep roots in far-right traditions, and longstanding anti-globalist, ruralist and environmental politics have been revitalised in the hope of capitalising on the popular sentiment and protest movements of the moment. In doing so, elements of the far right seek to justify their hatreds, and to redirect legitimate concerns for vital causes towards anti-minority sentiment.[41]

Mark Haus and Daniele Conversi describe it as a kind of *frame bridging* exercise where political actors seek to link ideologies that have similarities, although they are not structurally connected.[42] The example here, of course, is that of self-determination and environmentalism. They do this by picking up a 'theory of selective salvation propounded as a moral economy of "natural limits." This is set within an eschatological narrative of the preservation-rebirth of a hierarchical eternal order through a "great tribulation".'[43] Delf Rothe, in his discussion of the problem of the Anthropocene[44] in environmental justice movements and their associated literature, agrees.[45] He notes the difficulty in 'apocalyptic' movements and literature being captured by an eco-catastrophism that has roots in a Christian eschatology. As I have argued previously in this book, as well as in other places, this is an eschatology that looks only at *that which is to come*, rather than dwelling on the present and the need for demanding justice *now*. That is, these apocalyptic dialogues focus only on 'reducing presence of materiality and the present to the time of the now'.[46] This eco-catastrophism or green eschatology sees a future of global collapse and the exhaustion of natural resources, a kind of Malthusian future in which protectionism is the only option.[47] Bruno Latour has named a problem in the Western lexicon as the deification of the Anthropocene.[48] Anna Agathangelou, thinking with Latour, unpacks this by noting that what Latour helps us understand is that orders we have in place, that propel us forward in a linear manner, build a field of vision between order and disorder. Within this, some are meant to be the masters of nature, while others are threats to be vanquished. In this way, these 'eschatological' green nationalisms reproduce the imperial logics highlighted in the previous chapter in the discussions of Wynter, Fanon and the image of 'civilized man'. Aganthagelou further underlines this. In her article on denialism in Global Thought, she centres discussion on the knowledge production that I alluded to in Chapter 3 and argues that in their clinging to eschatological time, environmentalist movements and environmentalist literature are avoiding grappling with 'worldhoods whose trajectories are much longer and whose radical openness precedes' formations in the world of the present.[49] That is to say, indigenous knowledges and practices, long-term mobilizing by a variety of groups and communities, as well as intersectional struggles, are avoided, as are discussions of the exploitative aspects of world-building that have created our environmentally precarious present.

In reality, this does two things. The first is that it denies agency to the 'worldhoods' that have been around for much longer, subsuming these to the needs of an anxious West that must yet be the problem solver, the saviour from the catastrophe. This denies the mounting crises being faced by communities in the Global South as well as Indigenous communities everywhere because we are forced to a narrative that looks to protect an eventual problem of the Western world. A simple example was the continual speech acts at COP 26, where we heard leaders express their worry for their 'children, and their children's children', ignoring the arguments of the small island states and Indigenous leaders who asked for an urgent loss and damage response.[50]

The struggles of Indigenous communities, tribal groups and other marginalized communities are also a substantial, though not often acknowledged, part of this activism. These often involve mobilizations that occur on a variety of issues and multiple oppressions. This characteristic of the multiple is because justice is not something easily understood from texts or traditions – whether religious or not – and also because the meaning of justice varies considerably. Justice, liberation theologians will often remind us, is different in each context, and therefore our task is to understand reality from the horizon of a reconciled humanity. We then act accordingly to bring that reality about.

Quiet radicalisms in Canada

The Edmonton Outdoor Way of the Cross, held every year on a generally cold Good Friday in the city of Edmonton in Western Canada, is a very special event. The gathering is ecumenical and uses as its live stations different spaces in the city that work on aspects of social welfare, from homeless shelters to rehabilitation centres. The focus of this *Via Crucis* is to call attention to the links between faith and justice. As one of its organizers noted, what the event tries to do is to be people of faith who do not sit inside a church but look outside and suggest an understanding of what the world should be like.[51] There is a strong element here of the proclamation of Archbishop Romero, to be a church that unsettles, that proclaims, that gets under your skin.[52] Although attendance at the Way of the Cross is quite mixed, many of those who take part have been involved in social justice struggles for decades. Indeed, there is a kind of quiet radicalism to the social justice-minded Christians I met while living and working in Edmonton. Many are involved in the work of organizations such as Caritas Canada, Citizens for Public Justice and Kairos, or are members of, or frequent, the Marian Centre.[53]

A rainbow coalition

It is the word 'struggle' that marks out the activities taken up by various groups in the Global South – in this case South Asia – in the struggle for ecological justice. One of the key aspects of this has been the building of broad, regional coalitions such as the South Asian People's Action on Climate Crisis (SAPACC), the South Asia Conference on Environmental Justice, various interreligious networks across the subcontinent, South Asia Peasants coalition, Jubilee South and the South Asia Alliance for Poverty Eradication (SAAPE). What is important in many of these initiatives is that we see coming together farmers' organizations, trade union federations, Indigenous peoples' organizations, fisher groups, community groups working on gender issues, as well as socialist or generally left-leaning political parties. In Sri Lanka, for example, you may find an ecological justice coalition that brings together a radical Buddhist movement like the Sarvodaya movement, a national fisheries alliance and the Catholic Sisters of Charity. There is an understanding here, as with the post-Bandung theologies, that a radical break from traditional organizing, and a turn to alliance politics, is urgent and necessary. Activists involved have argued that, for the ecological justice struggle in South Asia to succeed, there must be a recognition that the climate crisis is a complex problem that requires a 'rainbow coalition' and new forms of strategizing.⁵⁴ Many advocates from South Asia also argue for the importance of broad coalitions against technocracy and market-centric agriculture that work instead to strengthen peasant agriculture across the region. Other scholars and activists agree with the need for such broad coalitions and point to the link between feminist struggles and ecological justice because it is often women who manage the local environment.⁵⁵ These coalitions, often centring on the activities and voices of peasant, indigenous, fisherpeople and farmers, have often been characterized as an environmentalism of the poor.⁵⁶ This, certainly, is because it is in forests, in the pollution of water bodies, in chemical agriculture and in coastal areas that the most significant impact of environmental degradation is felt. Others have written and continue to write brilliantly on these movements, so without further elaboration I will move to discussing the particular empirical case. However, it is important to note here that those religious women that I highlight choose to accompany these ecological justice struggles because of their theological understanding of a politics of struggle and the moral charge they feel to fight the oppressions imposed on peasant communities, indigenous, fisherpeople and farmers.

Involvement in the struggle has always been at the forefront for a significant portion of Sri Lanka's Christian community, especially its nuns and priests, to the extent that there is an entire section on Sri Lankan liberation theologians and Christian activists in South Asia in Open Archive's catalogue of dissidents and activists in Sri Lanka. I draw my reflections for this empirical case from the past four years of archival study of the liberation theology and Christian Marxist movements in Sri Lanka, as well as two years, between 2017 and 2019, spent observing the activities of Christian social justice activists in the country. Throughout Sri Lanka's post-independence history, Christian activists have not only joined popular movements, but have also been prolific in writing and publishing pamphlets, journals and newsletters on liberation, justice and human rights. Much of this, certainly, is connected to the Christian community's status as a marginal population in Sri Lanka, but there is also significant influence here from the imagination that took hold in the EATWOT years. I have never met a priest, nun or religious activist in Sri Lanka who did not exhibit a tremendous anger at the status quo, both at the institutional Church and at the government and global system of capital. In 2018, while observing a meeting of Sri Lankan EATWOT theologians, I was struck by a long sermon given by an older, habited Catholic nun who detailed for us the importance of the label Third World, and her sadness and anger that the term was now pejorative. As Paul Casperz wrote when reflecting on the role of the Christian Church in Sri Lanka, there was a conscious decision taken to work for a qualitatively new orientation in society, because the alternative would be stagnation and death, noting that any slight gain of the poor in Sri Lanka was often destroyed by price and production distortions.[57]

Scattered throughout the history of Christians in Sri Lanka is the activism of priests like Tissa Balasuriya and Michael Rodrigo (who was assassinated by the government) and organizations like the Christian Workers Fellowship and the Christian Solidarity Movement. The Catholic Sisters of Charity and the Nuns of the Holy Family, in particular, have maintained decades-long association with organizations such as the National Fisheries Solidarity Movement (NAFSO), or those working to secure rights for garment workers. Priests and nuns are also active in socialist parties like the Lanka Sama Samaja Party (LSSP) and have contributed their energy and their voices to movements against land grabs, against war, against militarization, and in seeking justice for the murdered and disappeared. The association with NAFSO has been a key part of the environmental justice work, as these are organizations that seek to protect the livelihoods of fishing communities. NAFSO, for example, is signed up to the Common Charter of the ESCR – a key focus of which is a shared critique of the dominant economic system. Nuns of the Holy Family who maintain a strong relationship with the women's federation of NAFSO draw directly on the ideas presented in Asian feminist liberation thinking to speak of their motivations for this activity, calling particular attention to the unique difficulties and persecutions that women face. The lending of the voices of religious leadership and visibly habited nuns has helped to make these frontline struggles stronger, and also make very explicit the rage against environmental destruction. A key point of contestation in most recent times has been the agitation to protect the country's wetlands and also to resist the Port City development project.[58] Asked why the Church engages in such struggles, one activist responded that 'the political authorities will sell the wetland for money but the Church raises its voice with the people, or we will not have an environment to live in the future'.[59]

The Port City has been a significant site of mobilization since the deal to build the city was signed in 2013 between the then Rajapaksa administration and a Chinese state-owned firm. The project was immediately seen as having ecologically adverse impacts on fisherpeople and land ownership, particularly due to planned granite extraction and land reclamation.[60] Indeed, in Colombo, where the Port City is being built, one can already feel the visceral heat in the area caused by the rapid reclamation of the sea and beach areas. The large amounts of sand dredged up by the reclamation work has also harmed traditional methods of fishing, which are usually undertaken in the shallower waters.[61] In December 2017 activists from the local Catholic church joined in the opposition being voiced by the local community and brought together a variety of social groups into the People's Movement against the Port City. Staging sit-ins and silent protests, and even enacting a *Via Crucis* through the streets of the capital, this group speaks of the death of the land, the sea, the fish and the community's livelihood, against those projects that, in the name of development, destroy Creation.[62] Thinking of the bleeding wounds of both environmental devastation and the impact on local communities, they speak of the loss of biodiversity and the loss of drinking water as crucial issues.

When asked for the motivations for their activity, female

activists I have spoken to, both lay and religious, draw direct links to the integral vision of the ecology embraced in the papal encyclical *Laudato Si'*, as well as the analytical understandings of Third World theologies. They see themselves in opposition to the lives lived by the elite of the world who, they say, will be the only people to profit from such a project. They see a prophetic role in speaking out against what they see as injustice and have no difficulty in linking the local to a global system of profit. They argue that land is sacred, a gift from God, and say that a situation where only a few are allowed to consume creates a situation that rips away the dignity of both the land and the people. The argument, then, is that it is time for the international community to take on the second kind of poverty, a humbling and a material lessening, and embrace the fact that the time has come to accept decreased growth. A Port City built on the dominant model of development that sees no immorality in unlimited production and the exploitation of natural resources is the certain opposite of this. The critique here, then, is of a Port City project that is able to *be* because of the 'normalisation of a global model that marginalises poor people ... what does this model do but pollute the air, the water, the soil?'[63] Since the 1960s, and now in the conjoining of their voices with ecological justice movements, these activists have been pointing out a need for a moral conversion, a radical break from unfettered growth which they mark out over and over again as Mammon, as what is evil.

Notes

1 Gerard Manley Hopkins, 2003, 'God's Grandeur', in John F. Thornton and Susan B. Varenne (eds), *Mortal Beauty, God's Grace: Major Poems and Spiritual Writings of Gerard Manley Hopkins*, New York: Vintage Books.

2 Kaleeg Hainsworth, 2014, *An Altar in the Wilderness*, Victoria, BC: Rocky Mountain Books.

3 Funded by the Louisville Institute: see Michael Agliardo, Sara Farid et al., '*Laudato Si'* Reception in North America: A Multi-City Study',

Louisville Institute, https://louisville-institute.org/our-impact/awards/collaborative-inquiry-team-discontinued/13088/, accessed 14.06.2022.

4 A. Sivanandan, 1982, *A Different Hunger: Writings on Black Resistance*, London: Pluto Press. See also Rohit Dasgupta, 2018, 'Rohit Dasgupta: Ambalavaner Sivanandan and Black Politics in Britain', *Theory, Culture and Society* (blog), 24 January, https://www.theoryculturesociety.org/blog/rohit-dasgupta-ambalavaner-sivanandan-black-politics-britain, accessed 14.06.2022.

5 Miriyam Aouragh, 2019, '"One Liberation Is Bound to Another": Beyond the Concept of White Privilege', *Race & Class* (blog), 1 October, https://irr.org.uk/article/one-liberation-is-bound-to-another-beyond-the-concept-of-white-privilege/; and also Miriyam Aouragh, '"White Privilege" and Shortcuts to Anti-Racism', *Race & Class* 61, no. 2 (October 2019), pp. 3–26, https://doi.org/10.1177/0306396819874629; Keeanga-Yamahtta Taylor (ed.), 2017, *How We Get Free: Black Feminism and the Combahee River Collective*, Chicago, IL: Haymarket Books; John Narayan, 'British Black Power: The Anti-Imperialism of Political Blackness and the Problem of Nativist Socialism', *The Sociological Review* 67, no. 5 (September 2019), pp. 945–67, https://doi.org/10.1177/0038026119845550.

6 Peter J. Smith et al. (eds), 2018, *The Role of Religion in Struggles for Global Justice: Faith in Justice?*, Rethinking Globalizations, London; New York: Routledge.

7 Frantz Fanon and Jean-Paul Sartre, 2001, *The Wretched of the Earth*, trans. Constance Farrington, reprint, Penguin Classics, London: Penguin Books.

8 Gary Haq and Alistair Paul, 2011, *Environmentalism since 1945*, London: Routledge, https://doi.org/10.4324/9780203803875.

9 Marco Armiero and Lise Sedrez (eds), 2014, *A History of Environmentalism: Local Struggles, Global Histories*, London: Bloomsbury Academic.

10 Gord Hill, 2009, *500 Years of Indigenous Resistance*, Oakland, CA: PM Press.

11 Eve Tuck and K. Wayne Yang, 'Decolonization Is Not a Metaphor', *Decolonisation: Indigeneity and Society* 1, no. 1 (2012), https://jps.library.utoronto.ca/index.php/des/article/view/18630, accessed 14.06.2022.

12 A selection of examples includes but is not limited to: Christopher Rootes (ed.), 2013, *Acting Locally: Local Environmental Mobilizations and Campaigns*, London: Routledge; Kevin Dunion, 2012, *Troublemakers: The Struggle for Environmental Justice in Scotland*, Edinburgh: Edinburgh University Press; Ryan Holifield, Jayajit Chakraborty and Gordon Walker, 2018, *The Routledge Handbook of Environmental Justice*, Routledge International Handbooks, Abingdon: Rout-

ledge; Dorceta E. Taylor, 'The Rise of the Environmental Justice Paradigm: Injustice Framing and the Social Construction of Environmental Discourses', *American Behavioral Scientist* 43, no. 4 (January 2000), pp. 508–80, https://doi.org/10.1177/0002764200043004003; Dorceta E. Taylor, 'Introduction: The Evolution of Environmental Justice Activism, Research, and Scholarship', *Environmental Practice* 13, no. 4 (December 2011), pp. 280–301, https://doi.org/10.1017/S1466046611000329; Michael Méndez, 2020, *Climate Change from the Streets: How Conflict and Collaboration Strengthen the Environmental Justice Movement*, New Haven, CT: Yale University Press.

13 Although I make a differentiation between faith-based and secular for the sake of brevity, I do this with the awareness that religion and religious actors often play diverse roles and that it is difficult to clearly distinguish between religious and secular. This is especially true because religion is deeply intertwined in political mobilizations for justice, especially at local levels.

14 Donatella Porta and Dieter Rucht, 'The Dynamics of Environmental Campaigns', *Mobilization: An International Quarterly* 7, no. 1 (1 February 2002), pp. 1–14.

15 Willis Jenkins, Evan Berry and Luke Beck Kreider, 'Religion and Climate Change', *Annual Review of Environment and Resources* 43, no. 1 (17 October 2018), pp. 85–108, https://doi.org/10.1146/annurev-environ-102017-025855.

16 Anupama Ranawana, 'Covid-19, Localisation and the Role of the Faith Actor', London: Christian Aid, November 2021; Joanne M. Moyer, A. John Sinclair and Harry Spaling, 'Working for God and Sustainability: The Activities of Faith-Based Organizations in Kenya', *VOLUNTAS: International Journal of Voluntary and Nonprofit Organizations* 23, no. 4 (1 December 2012), pp. 959–92, https://doi.org/10.1007/s11266-011-9245-x; Olivia Wilkinson, 2021, '"As local as possible, as international as necessary." Investigating the place of religious and faith-based actors in the localization of the international humanitarian system', in Jayeel Cornelio et al. (eds), *Routledge International Handbook of Religion in Global Society*, Routledge International Handbooks, New York: Routledge.

17 JLI Faith and Communities, 2019, 'JLI May 21, 2019, Webinar on Faith-Based Climate Programs', *YouTube*, 21 May, https://www.youtube.com/watch?v=KcgWCE5v-WI, accessed 15.06.2022. Dr Olivia Wilkinson, the JLI Director of Research, led and moderated the webinar.

18 Kahsay Haile Zemo and Halefom Yigzaw Nigus, 'Does Religion Promote Pro-Environmental Behaviour? A Cross-Country Investigation', *Journal of Environmental Economics and Policy* 10, no. 1 (2 January 2021), pp. 90–113, https://doi.org/10.1080/21606544.2020.1796820.

19 Robin Globus Veldman, 2016, *How the World's Religions Are*

Responding to Climate Change: Social Scientific Investigations, Abingdon: Routledge.

20 Dalai Lama, 'A Buddhist Concept of Nature' (New Delhi, February 1992), *His Holiness the 14th Dalai Lama of Tibet*, https://www.dalailama.com/messages/environment/buddhist-concept-of-nature, accessed 15.06.2022. The Dalai Lama ends this statement with an exhortation that working to protect and preserve the environment begins with the human heart. This runs in parallel with the urgings of Pope Francis and Patriarch Bartholomew towards an ecological conversion. From faith leaders we see that what is asked for is a moral awakening, even a moral unsettling.

21 Katie Javanaud, 'The World on Fire: A Buddhist Response to the Environmental Crisis', *Religions* 11, no. 8 (2020), https://doi.org/10.3390/rel11080381.

22 John Stanley, David Loy and Gyurme Dorje (eds), 2009, *A Buddhist Response to the Climate Emergency*, Boston, MA: Wisdom Publications.

23 William Avis, 'Role of Faith and Belief in Environmental Engagement and Action in MENA Region' (University of Birmingham: The K4D helpdesk service, 19 May 2021).

24 Ahmed Afzaal, 'Disenchantment and the Environmental Crisis: Lynn White Jr, Max Weber, and Muhammad Iqbal', *Worldviews: Global Religions, Culture, and Ecology* 16 (1 January 2012), pp. 239–62, https://doi.org/10.1163/15685357-01603004.

25 'A Rabbinic Letter on the Climate Crisis', 2015, *The Shalom Center*, 12 May, https://theshalomcenter.org/civicrm/petition/sign?sid=17, accessed 15.06.2022.

26 David Krantz, 'Shmita Revolution: The Reclamation and Reinvention of the Sabbatical Year', *Religions* 7, no. 8 (2016), https://doi.org/10.3390/rel7080100.

27 Joelle Jackson, 2021, 'Repairing Our World: Jewish Environmentalism through Text, Tradition, and Activism', *Folklife* (blog), 12 January, https://folklife.si.edu/magazine/jewish-environmentalism-text-tradition-activism, accessed 15.06.2022.

28 Zoe S. Todd, 2015, 'Indigenizing the Anthropocene', in Heather Davis and Etienne Turpin (eds), *Art in the Anthropocene: Encounters among Aesthetics, Politics, Environments and Epistemologies*, 1st edn, Critical Climate Change, London: Open Humanities Press.

29 Andreas Malm and Alf Hornborg, 'The Geology of Mankind? A Critique of the Anthropocene Narrative', *The Anthropocene Review* 1, no. 1 (1 April 2014), pp. 62–9, https://doi.org/10.1177/2053019613516291.

30 Shekhar Pathak and Manisha Chaudhry, 2021, *The Chipko*

Movement: A People's History, Ranikhet: Permanent Black in association with Ashoka University, Bangalore.

31 Vandana Shiva, 'The Tree Saviours of Chipko Andolan. A Woman-Led Movement in India', *KOSMOS: Journal for Global Transformation* (Autumn 2020), https://www.kosmosjournal.org/kj_article/99176/, accessed 15.06.2022.

32 As the purpose of this chapter is to focus more on the empirical cases, I have not presented here all the different arguments and discussions within the quite vast field of religion and climate change. However, many of the works referenced in this chapter, especially the work of Willis Jenkins, Randy Haluza-Delay, Linda Woodhead and others, provide excellent scholarship on these debates, as well as on the different evolutions of the relationship between religion and the struggle for environmental justice.

33 James Trafford and Anupama Ranawana, 'Imperialist Environmentalism and Decolonial Struggle', *Discover Society* 71 (August 2019), https://archive.discoversociety.org/2019/08/07/imperialist-environ mentalism-and-decolonial-struggle/, accessed 15.06.2022.

34 James Trafford, 2021, *The Empire at Home: Internal Colonies and the End of Britain*, London: Pluto Press.

35 Due to the need to focus on different aspects, I will not spend time discussing other problematic issues of XR such as their embrace of being arrested – tactics which immediately exclude persons of colour and immigrants from taking part in mobilizations and public demonstrations. Should I, a person on a visa, take part in such a demonstration and allow myself to be arrested, I would be deported. Bell and Bevan's study into BAME and working-class communities' perceptions of XR is illuminating here: Karen Bell and Gnisha Bevan, 'Beyond Inclusion? Perceptions of the Extent to Which Extinction Rebellion Speaks to, and for, Black, Asian and Minority Ethnic (BAME) and Working-Class Communities', *Local Environment* 26, no. 10 (3 October 2021), pp. 1205–20, https://doi.org/10.1080/13549839.2021.1970728. See also this BBC report on the lack of diversity in environmental justice movements in the UK: Hanna Yusuf, 2019, 'Climate activism failing to represent BAME groups, say campaigners', *BBC News*, 27 May, https://www.bbc.co.uk/news/uk-48373540, accessed 15.06.2022.

36 Rupert Read, 'Love Immigrants, Rather than Large-Scale Immigration', *The Ecologist*, June 2014.

37 I defer to Clifford Leeks' definition of whiteness as 'as a set of practices that function to protect and maintain privilege, while others define whiteness simply as the experience of privilege'. This being said, I think it also important to unpack and understand what is meant by these terms. Although this is not the subject of this book, I refer to this

excellent essay by the philosopher Helen Ngo that provides important discussion points: https://overland.org.au/2020/06/on-white-privilege-white-priority-and-white-supremacy/, accessed 15.06.2022.

38 Wretched of the Earth Collective, 2019, 'An Open Letter to Extinction Rebellion', *Red Pepper UK* (blog), 3 May, https://www.redpepper.org.uk/an-open-letter-to-extinction-rebellion/, accessed 15.06.2022.

39 Norimitsu Onishi, 2017, 'France's Far Right Wants to Be an Environmental Party, Too', *New York Times*, May, https://www.nytimes.com/2019/10/17/world/europe/france-far-right-environment.html, accessed 15.06.2022.

40 TRT World, 2021, 'Far-Right Ideas Have Grown "Exponentially" in the UK', *TRT World*, 26 March, https://www.trtworld.com/magazine/far-right-ideas-have-grown-exponentially-in-the-uk-45342, accessed 15.06.2022.

41 Nick Lowles and Nick Ryan (eds), 2021, 'State of Hate 2021 Backlash, Conspiracies & Confrontation'; London: Hope not Hate, https://hopenothate.org.uk/wp-content/uploads/2021/03/state-of-hate-2021-final-2.pdf, accessed 15.06.2022.

42 Daniele Conversi and Mark Friis Hau, 'Green Nationalism. Climate Action and Environmentalism in Left Nationalist Parties', *Environmental Politics* 30, no. 7 (10 November 2021), pp. 1089–110, https://doi.org/10.1080/09644016.2021.1907096.

43 Angela Mitropoulos, 'Lifeboat Capitalism, Catastrophism, Borders', *Dispatches* 162, no. 000 (2018), http://dispatchesjournal.org/articles/162/, accessed 15.06.2022.

44 I use here the definitions and descriptions provided by Meera Subramanian, Jan Zalasiewicz et al.:

> The *term* Anthropocene initially emerged from the Earth System science community in the early 2000s, denoting a *concept* that the Holocene Epoch has terminated as a consequence of human activities. First associated with the onset of the Industrial Revolution, it was then more closely linked with the Great Acceleration in industrialization and globalization from the 1950s that fundamentally modified physical, chemical, and biological signals in geological archives. This reflects an Earth System state in which human activities have become predominant drivers of modifications to the stratigraphic record, making it clearly distinct from the Holocene. However, more recently, the term Anthropocene has also become used for different conceptual interpretations in diverse scholarly fields, including the environmental and social sciences and humanities. These broader conceptualizations encompass wide ranges and levels of human impacts and interactions with the environment.

Writing as I am from within the humanities, my understanding of the Anthropocene is then to think within the boundaries of such human impacts, interactions and discourses. Jan Zalasiewicz et al., 'The Anthropocene: Comparing Its Meaning in Geology (Chronostratigraphy) with Conceptual Approaches Arising in Other Disciplines', *Earth's Future* 9, no. 3 (March 2021), https://doi.org/10.1029/2020EF001896. Meera Subramanian, 'Humans versus Earth: The quest to define the Anthropocene', *Nature* 572, no. 7768 (August 2019), pp. 168–70, https://doi.org/10.1038/d41586-019-02381-2.

45 Delf Rothe, 'Governing the End Times? Planet Politics and the Secular Eschatology of the Anthropocene', *Millennium* 48, no. 2 (1 January 2020), pp. 143–64, https://doi.org/10.1177/0305829819889138.

46 Rolando Vazquez, 'Precedence, Earth and the Anthropocene: Decolonizing Design', *Design Philosophy Papers* 15, no. 1 (2 January 2017), pp. 77–91, https://doi.org/10.1080/14487136.2017.1303130.

47 In his 1798 'Essay on the Principle of Population', the theologian and economist Thomas Malthus described a vicious cycle between increased food production, population growth, resource depletion and the resulting spread of famines, disease and moral deformation among the lower classes. Malthus's work has inspired others, like Ehlrich, to ask questions regarding 'carrying capacity' to understand the maximum human population the earth can sustain. Such questions very easily lend themselves to imperialistic notions of 'resource and population management'. They also look only at an 'end times' rather than at the crisis occurring in the Global South and in impoverished communities within history and in the present.

48 The University of Edinburgh, 2013, 'Prof. Bruno Latour - The Anthropocene and the Destruction of the Image of the Globe', *Youtube*, 1 March, https://www.youtube.com/watch?v=4-l6FQN4P1c, accessed 15.06.2022. Professor Bruno Latour delivers the Gifford Lecture series entitled 'Facing Gaia. A new enquiry into Natural Religion'.

49 Anna M. Agathangelou and ProtoSociology Project, 'Real Leaps in the Times of the Anthropocene: Failure and Denial and "Global" Thought', *ProtoSociology* 33 (2016), pp. 58–92, https://doi.org/10.5840/protosociology2016334.

50 BBC, 2021, 'COP26: "Two Degrees Is a Death Sentence" for Island Nations – Barbados PM Mia Mottley', *BBC News*, 1 November, https://www.bbc.com/news/av/world-59117750, accessed 15.06.2022.

51 Jonny Wakefiled, 2019, 'Hundreds of Faithful Gather for Way of the Cross', *Edmonton Journal*, 19 April, https://edmontonjournal.com/news/local-news/hundreds-of-faithful-gather-for-way-of-the-cross, accessed 15.06.2022.

52 Óscar A. Romero and James R. Brockman, 2004, *The Violence of Love*, Maryknoll, NY: Orbis Books, 2004.

53 The Marian Centre is part of the Madonna House Apostolate, founded by the writer and activist Catherine Doherty. A Russian émigré, Doherty, much like her frequent correspondent Dorothy Day, lived a 'radical gospel' and soon founded Friendship House in a poor part of Toronto, Ontario. Friendship House soon became a shelter for the homeless, provided meals for the hungry, recreation and books for the young and a newspaper to make known the social teachings of the Church. In 1938, Catherine initiated an interracial apostolate in Harlem, New York, living with and serving the African Americans. After some practical setbacks Doherty then moved to the north-east of Toronto to continue her work, founding the Madonna House community, an open family of lay men, lay women and priests, living in love and breathing from the 'two lungs', East and West, of the Catholic Church. A key read on Doherty is: Catherine de Hueck Doherty, Marian Heiberger, 2002, *Living the Gospel without Compromise*, Combermere, ON: Madonna House Publications.

54 Nagraj Adve, 2020, 'South Asian Coalition Links Climate Demands with Social Struggles', *People and Nature*, 21 February, https://peopleandnature.wordpress.com/?s=south+asian+coalition, accessed 15.06.2022.

55 Buddhima Padmasiri and Samanthi Gunawardana, 2021, 'Rural Women's Resistance to Neoliberal Agricultural Reform: The Women of Monaragala Sri Lanka', *Progress in Political Economy*, 8 July, https://www.ppesydney.net/rural-womens-resistance-to-neoliberal-agricultural-reform-the-women-of-monaragala-sri-lanka/, accessed 15.06.2022. Alyssa Banford and Cameron Kiely Froude, 'Ecofeminism and Natural Disasters: Sri Lankan Women Post-Tsunami', *Journal of International Women's Studies* 16 (2015), pp. 170–87. For a broader study on women at the frontlines of climate change see: Nushat Chowdhry et al., 2021, 'Women on the Front Line Healing the Earth, Seeking Justice', London: Christian Aid, October, https://www.christianaid.org.uk/sites/default/files/2021-10/Women%20on%20the%20Front%20Line%20Final%20report.pdf, accessed 15.06.2022.

56 P. Basu, 'Environmental Justice in South and Southeast Asia: Inequalities and Struggles in Rural and Urban Contexts', in Ryan Holifield (ed.), *The Routledge Handbook of Environmental Justice*, London; New York: Routledge.

57 Paul Caspersz, 'The Role of the Christian Church in Contemporary Sri Lanka', *Vidyajyoti – Journal of Theological Reflection* 53 (June 1989), pp. 289–98.

58 Environmental Justice Atlas, 2018, 'Fisher Folks, Environmentalists and Religious Leaders against the Colombo Port City, Sri Lanka', *Environmental Justice Atlas* (blog) (May), https://ejatlas.org/conflict/

fisherwomens-mobilization-against-the-port-city-sri-lanka, accessed 15.06.2022.

59 UCA News, 2021, 'Priests, Nuns Oppose Environmental Destruction in Sri Lanka', 16 July, https://www.facebook.com/watch/?v=269749761621132, accessed 15.06.2022.

60 Vinitha Revi, 2021, 'Colombo Port City Project: Controversial since its inception', *Observer Research Foundation*, 28 December, https://www.orfonline.org/expert-speak/colombo-port-city-project/#:~:text=There%20are%20numerous%20related%20issues,particularly%20the%20famous%20Negombo%20lagoon, accessed 15.06.2022.

61 Asanga Guwanansa, no date, 'Creation of New Urban Land by Reclaiming the Sea in Colombo Port City, Sri Lanka', UN Habitat, https://unhabitat.org/sites/default/files/download-manager-files/1544097120wpdm_Colombo%20Case%20Study.pdf, accessed 15.06.2022.

62 Melani Manel Perera, 2016, 'Colombo: Via Crucis of Catholics and Fishermen against Port City Project', *Asia News*, 24 March.

63 Melani Manel Perera, 2016, 'Catholics, Fishermen and Environmentalists Protest Reopening of Colombo Port City', *Asia News*, 9 January.

5

A Theology of Rage

How do we bear this knowledge?[1]
Be angry (Ephesians 4.26)

The outpourings of rage following the death of George Floyd and the rage-filled demonstrations at the last few COP events make it hard to turn our eyes away from how rage is productive and is a key element of movements towards justice. We also cannot ignore the rage that hurtles towards us from this Earth that we have almost casually brutalized. This dying world is angry. There is a furiousness to the waves of water that drown entire towns, to the heat that bakes the earth of islands and denies life. Creation groans, shudders and shakes with fury. Living now in the north-east of Scotland, with the constant increase in stormy weather, I feel the pain and fury of the created world on my body as 50 mile an hour winds whip at my face and my five-foot frame as I struggle up a hill. An ocean away, the fury of the monsoon in Karnataka derails trains and brings the earth down in landslides in several townships, killing a number of people.[2] Even the Secretary-General of the United Nations refers to the climate crisis as nature striking back with fury.[3] Repairing this world, and responding to this fury and this shaking, also requires a furious response. In previous chapters I have tried to recount what is being said in theology, in the long history of church documents on the environment, on the importance of ecological sin and ecological conversion, on the histories of brutality and exploitation and, finally, the demands for justice from activists and communities who are living at the mercy of a dying, angry world. There is little that has been recounted that should be new, or necessarily revelatory. These stories, these needs, are part of the history of our social world; what is necessary is that we pay attention to, and enact, the spiritual and moral awakening that is required. Rage also brings together conjoined struggles, thus linking the work for climate justice to the work for racial justice.

I began to think about the need to speak more explicitly about rage while doing the rounds at different events on faith and climate change. Each of these were excellent events, with brilliant speakers and active, engaged audiences. Most of the audiences were young and situated in the Global North. We spoke about grief, and sadness, and the end of the world. We rarely spoke – unless the speaker was from an Indigenous or global minority background – of the fact that many places around the world are facing the climate crisis *now*. Yes, there was the debate about the Global South being more severely impacted, but having this discussion without understanding the need for systematic change only brings the Global North back into the centre and continues framing the Global South as being without agency. Over and over again we were asked, 'What can I do? What can I change about my life?', 'How do I deal with my grief, how can we lament?' I wondered what it was that was being lamented. I was also struck by the privilege of being able to grieve – for the world as it dies, for the loss of lifestyles that we live – while humans and animals in other parts of the world were quite literally fighting for their lives. It is not only an anxiousness about what is to come, but also a worry over what needs to be 'let go'. I feel angry every time I am asked what needs to be done, or asked how we can gather to lament together, and lately my response has been that of Christina Sharpe, to remember that for so very many there is no choice of lament.[4] I am not saying that it is immoral to grieve, to be anxious or to lament the climate crisis that we find ourselves in. As Panu Pihkala has argued, climate change feels like death.[5] Ecological grief is part of a pantheon of emotions related to fear and worry, even a kind of practical anxiety, within the scope of emotions surrounding the environmental crisis.[6]

Climate grief is related both to changes that have already happened and to changes that are coming or are in the process of happening. Thus, climate grief often has elements of what the grief theorists call 'anticipatory grief' or 'transitional grief'. This complicates things. All kinds of grieving can be difficult in our contemporary societies, where skills in private or public grieving have long been neglected. But anticipatory grief is always hard. What is truly lost, or will be? When do we grieve those losses – when they begin, or when they end? It is easier to grieve concrete losses that have happened, such as the loss of a certain part of ecosystem or community. But how does one grieve losses which are ongoing for decades?[7]

Grief is an entirely valid response, and those who are experiencing, and who have been at the receiving end of, histories of extraction and exploitation do indeed grieve. Grief and lament are deeply theological spaces. As Hannah Malcolm notes in her introduction to *Words for a Dying World*, the work of this theological grief is an attempt to navigate and articulate the great trauma of having the relationships between human and non-human rupture.[8] 'We grieve', she says, 'the death of particular things.' And we grieve, like Malcolm and many others who write on ecological grief and hope, because we love and are moved by that love.[9] The intervention I wish to provide here is that grief suggests time and space. In saying this, what I hope we can acknowledge is that there is privilege to having the time and space to express grief.[10] Time is not a resource that is easily found or accessed. In order to be able to have this time and space, we must be in a physically possible position that allows for this outpouring of emotion and the processing of it. If we are able to find the time to lament loss, this also indicates that we are in a position to do so, and that we have much to lose. Some of the activists with whom we journeyed in Chapter 4 would often point out that there is a distinct privilege in being able to experience grief and be in a space where one can sit with an emotion. These activists and communities are very clear in pointing out that one of the reasons for the mobilizations that they take part in is the decades of experiencing and navigating climate grief. This grief is turned into collectivizing and broad alliances because it is understood that grief doesn't move us forward. Indigenous communities, persons of colour and communities of the Global South have not only experienced brutalism, but continue to do so. As the academic Farhana Sultana declares, 'We do not have the luxury of wallowing in our suffering ... we resist, survive and collectivize.'[11] For the activists and communities in the Global South and in Indigenous spaces, acting on climate issues is also about necessary survival within a system that does not want these individuals to live.

We must acknowledge that those who have faced histories of oppression and extraction may already feel that their lives have no meaning or worth and, as such, they may find no need to grieve. Sayanti Ghosh, looking at the hierarchy of death in the aftermath of the first Covid-19 wave in India, points out how bodies that have become historically used to oppression might argue that life has no meaning when you live in a state of precariousness, and when you know that your life and way of living are meaningless to others.[12] 'How meaningful is their death, how non-existent is their grief?' she asks. Nylah Burton, writing on climate grief, would concur, noting that people of colour experience climate grief alongside long histories connected to the pain and suffering of living always with racial terror, of knowing already that one's life means less.[13] Pihkala asks, in the paragraph quoted above, what have we lost, and how do we grieve? I ask again, what is it we are grieving for, even if we grieve in and through love? This same question of what is it that we grieve is also asked by Shelly Brewster in her discussion of eco-anxiety and performative acts of mourning.[14] Brewster argues that there is an element of 'white tears' that can then be used to avoid the structural causes of environmental violence. Grief is then platformed as an action, but it may not necessarily be a transformative action. In fact, one could argue that grief may lead only to a kind of navel-gazing, an inward and paralysing emotion that is also then coupled with feeling burdened by one's complicity in the histories and presents of extraction. Sarah Jaquette Ray would agree, arguing strongly that there is an inherent whiteness to climate anxiety.[15] She asks if there is a kind of coding in many of the conversations surrounding climate grief that is also related to anxiety over losing the comfort of privilege. Ray, as well as others, point out that this anxiety is also increasingly linked to seeing climate refugees as threats, a subject that has been covered extensively in this book in Chapter 4's discussion of green nationalism.[16] It is a short logical step from fear of the apocalypse to fear of the horde. In fact, the key exploration of Ray's work is to underline how early environmental activists also had close links to, or were, anti-immigrant eugenicists.[17] Is the intensity of climate grief transformative? My response would be that the work for transformation and the moral awakening that is required cannot be grief, because grief turns too easily inward and into the self. This looking inward, even perhaps the wallowing in grief, means that it is then easy to turn our hearts and minds away from justice, to insert apocalypse into the narrative of ecological justice and not hopeful, radical justice. As I noted in the introductory chapter, the justice narrative is key to our moral and spiritual awakening. We must look at the history and present of the world and feel prophetic rage. The climate crisis is not a new crisis, and we must come together in rage to overturn systems of enslavement, expropriation, colonization and indigenous genocide. Without feeling rage or acknowledging and centring the rage of those who have been oppressed and exploited, we cannot move forward. This rage, poured out, can be mobilizing.[18] Our theological and pastoral spaces must not be afraid of rage.

Theologies of rage

There are, of course, many scriptural sources for righteous anger, not least Jesus Christ's many *thunderings* to his followers. Christ's righteous anger is one that is grieved by sin, as we see in Mark 3.5, Matthew 21.12, Mark 3.1–6 and several other instances.[19] Anger continues as a theme of the Good News of Christ and I have used a quote from Ephesians at the top of this chapter. In doing so, I am also referencing Amy Allen's work on good anger in the New Testament.[20] Such anger is important because it seeks to critique and reform, and also because it stands with those who are most oppressed. Anger as articulated in the passage in Ephesians is also anger that recognizes what is sinful and destroys it. Vincent Lloyd, thinking very systematically with Scripture, also argues that we must stop thinking of anger in a secularized manner.[21] Indeed, Lloyd agrees with me that when we see the anger articulated by voices within oppressed communities, what we find is prophetic anger. Lloyd in fact argues that attending to the anger of these communities is practical theological work and calls us to discern with and in anger. Doing so is part of the journey of conversion. In my reading of anger, this is an important aspect of the journey into ecological conversion. That is, to recognize and foreground the prophetic anger of the margins and let such rage effect one's conversion of heart and mind.

I pick up the concept of rage because it is the overwhelming emotion that I notice when I have conversations with activists, Earth defenders, religious communities and scholars in the Global South. As the previous chapters, as well as other theologians, have argued, we must be aware of the different dimensions of anger present within these oppressed communities. This is not only because they recognize and foreground their agency as leaders in social movements, but also because such anger provides deep insight, thus informing our theologies of justice. To reference a central tenet of liberation theology, justice is different and varied across communities and there is a need to embrace this multidimensional aspect.[22] It is often articulated in prophetic rage. As noted in the Introduction and then further elaborated upon in Chapters 3 and 4, anti-colonial resistance has long been interwoven with environmental protection and we can see that there is a generative rage that binds these struggles together, a kind of rainbow coalition. It is the overwhelming epistemic emotion that runs through liberation theology, especially the liberational theologies of the Third World as they argue for epistemological breaks from their Western counterparts and raise up the problem of justice.[23] As I have spent significant time on theologies of the Third World in previous chapters, I will not continue to elaborate here. However, I will say that it is often easier to recognize Third World voices or oppressed communities only when they are 'good' – that is, when they present themselves gently and in silent gratefulness, negating their actual experience of anger; perhaps because of this we may not always read liberative theologies and see the rage that drives them. Joshua Samuel, for example, argues that anger is an essential feature of liberation theologies, especially those that write from Black and Dalit perspectives.[24] He notes that Dalit theologies are inspired by the cross because it is a symbol of revolution that asks for a rising up against oppression, and therefore they cannot be genuine without the emotion of anger. Thus Dalit theologians are connecting with the Christ who came to set fire to the earth.[25]

This is not to say that these theologies are not motivated by love, or the experience of Divine Love. Being moved by love, and having encountered the Being that is loving, are we not moved to anger to see the lack of love that has been expended upon the Earth? Gutiérrez, for example, always remarks that the real question of loving faith is how we are able to speak about God in this world that is suffering.[26] Surely the answer must also lie in rage. Earlier on in this book, I quoted Pope Francis' exhortation to feel outrage as Moses did, as Christ did, and even as the Creator does in the face of injustice. The sentences that follow the exhortation appeal to our ability to love fiercely:

> The incidents of injustice and cruelty that took place in the Amazon region even in the last century ought to provoke profound abhorrence, but they should also make us more sensitive to the need to acknowledge current forms of human exploitation, abuse and killing.[27]

I am not the first theologian to talk about the importance of rage. There is, in fact, a long line of prophetic theological thought that sees anger as important in speaking truth to power. Throughout several Christian traditions, there have been many voices arguing that anger is a virtue.[28] Feminist theologians, both from Western and non-Western spaces, have challenged the conventional understanding of anger as a deadly sin and highlighted its virtuous aspect. Rosemary Radford Ruether identifies anger alongside pride with a theological virtue that is 'grace welling up from the ground' of the oppressed women to 'transcend false consciousness and break its chains'.[29] Indeed, James Cone once remarked that his first book, *Black Theology, Black Power*, was written in a moment of rage.[30] The text was written shortly after the death of Martin Luther King Jr and Cone raises this sense of righteous indignation and turns it into theology that is not only prolific but productive and mobilizing. Joshua Lazard, reading Cone, argues that if we do not hold space for such anger, we are also denying the spiritual encounter of our righteous indignation.[31] For Lazard, writing about Black anger elevates and fulfils the Black Christian experience as it struggles to resist racist violence. Xolani Kacela agrees, noting that an explicitly articulated anger can be a constructive outlet for organizing the anti-racist struggle.[32] For both Lazard and Cone, a theology of anger also helps to discern the next move of the revolution. 'If the Church is to remain faithful to its Lord ... It must become prophetic, demanding a radical change in the interlocking structures of this society.'[33]

Johnny Bernard Hill's work *Prophetic Rage: A Postcolonial Theology of Rage* continues this argument in a full treatment, noting that the ability to resist imperialism and to create spaces of peace and justice requires prophetic rage.[34] Hill sees such rage as power and courage, noting examples from within the American civil rights movement. Hill's argument is similar to mine in that he sees the prospect of world transformation as a project of resisting empire. This requires mobilization and solidarity, and such collectivizing can best come from a space of rage. Hill also sees a connection between rage and grief, where the latter is the first step towards the former, 'the kind of theological praxis expressed in the life and words of the ancient prophets and the prophetic witnesses of the church to engender a just society'.[35] To think in and with prophetic rage is also to foreground the voices of those who have experienced oppression, as argued above and also in C. L. Nash's searing work on womanist prophetic rage.[36] Nash, noting the experience of women who encountered the Divine even in the midst of slavery, argues that such prophetic rage transforms our epistemes. Such prophetic rage leads us into community. Felix Wilfred, thinking from both the Christian and Buddhist traditions, agrees, noting that rage understood prophetically is a way to jolt people out of a sense of apathy or indifference.[37] Certainly, some of my own and others' arguments given above are directed towards the fact that grief can be a way of looking only at one's own anxiety and turning away from the suffering of others. Rage, especially when channelled prophetically, can challenge such inwardness because we are then charged with our human responsibility, our charge as persons of faith to be dismayed at how the beauty in Creation has been distorted.

Creative, world-changing rage

I could perhaps write another book that explores the different theologies of anger or, shall we say, how the theologies of liberation are angry ones. What I wished to point out in the section above are just brief examples of how this idea of a theology of rage is not an innovation that I bring to the table, but simply a tradition I wish to build upon and especially to extend to the discussions surrounding ecological justice. I use the phrase 'prophetic anger' because what I am linking to, as I have previously mentioned, is what is articulated by the activists and communities highlighted in Chapter 4, and the just world that they and liberative theologies demand. This is an understanding of an anger that is deep, almost illegible, and therefore it is not your normal, everyday anger, but anger that seeks to change the world.[38] We can use this rage strategically to build an 'oppositional culture', that is, a social model shaped to symbolize an explicit alternative to the prevailing dominant socio-political order. Maria Lugones calls this a kind of separatist anger that is experienced differently and causes one to stand apart and refuse the normative anger that recognizes the need for justice and works towards it.[39] This is why Lloyd sees this as anger in a theological idiom.[40] I would argue that we need more theological work, not only from a liberation perspective, that considers how to theologize rage and harness it as a resource, and how scholarship requires rage to make epistemological breaks. More than this, seeing the rage that charges the everyday agitations of those living in oppressed and exploited communities, we must try to enter into this rage as a necessary act of conversion.

Working for justice is exhausting, but it cannot cease. The energy for the work of justice must come from somewhere, and certainly it can come from the kind of prophetic rage described above, because it is rage that sees injustice and cannot bear the desecration of the created world. Grief can be a wearying business, but grief and anger woven together are what fuels the movement for justice. Prophetic rage challenges and organizes because it sees not only what needs to be transformed but that it can thus be transformed. Natalia Imperiatori-Lee, reviewing ethnographic work on the working-class community and women in America, notes that anger has been fruitful and generative for these social movements, because the movement has been born out of the anger of the marginalized.[41] Such rage can be used for the productive activity of building resistance and therefore informing a liberating perspective. It is what the philosopher Myisha Cherry calls 'Lordean rage', referencing the anti-imperial work of the writer Audre Lorde.[42] Lorde argued that rage, focused with precision, was world-changing and world-making.[43] Lorde's work often reminded us of the need for rage, because so very many communities were not meant to survive the European project of colonialism. Part of the rage we must feel is about undoing this colonial project. Rage is not only about the liberation of those who are oppressed; it is also about bringing about reconciliation between the oppressed and the oppressors. Without this kind of rage structuring our ecological conversion, we will lose the justice narrative, and we will also be too fatigued to continue our work for justice. As I argued in the Introduction, without a justice narrative at the heart of our ecological work, we simply become vehicles for the privileged, and can even find ourselves moving in the direction of a racist, eugenicist green nationalism. Working for ecological justice, if not inspired by the rage of those who have been oppressed, will become further complicit in neo-colonialism and continued oppression in the imperialist core. We cannot embrace an electric car world if we are not aware of what lands are mined and what peoples are displaced in the search for the lithium for the car battery.

Let us make rage central so that we can begin to think and act differently. Let us refuse destruction and, as I argued when I wrote my introduction a year ago, pour back in love, in grief, and also in rage. We must have restitution and reparation. In order to do this, we must be affected by the rage of others, and feel this rage itself as a necessary first step towards recognizing our ecological sin, repenting of it, and truly undergoing our ecological conversion.

Notes

1 Michele Lobo et al., 2021, 'Earth Unbound: Climate Change, Activism and Justice', *Educational Philosophy and Theory* 53, no. 14 (6 December), pp. 1491–508, https://doi.org/10.1080/00131857.2020.1866541.

2 Karnataka Bureau, 2021, 'Monsoon Fury Throws Life out of Gear in Karnataka', *The Hindu*, 24 July, https://www.thehindu.com/news/national/karnataka/monsoon-fury-throws-life-out-of-gear-in-karnataka/article35503342.ece, accessed 15.06.2022.

3 Antonio Guterres, 2019, 'Nature Striking Back with Fury, Secretary-General Tells Climate Action Summit, Calling "Apocalyptic" Bahamas Destruction "the Future – If We Do Not Act Now"',

United Nations, https://www.un.org/press/en/2019/sgsm19757.doc.htm, accessed 15.06.2022.

4 Christina Elizabeth Sharpe, 2016, *In the Wake: On Blackness and Being*, Durham, NC: Duke University Press.

5 Panu Pihkala, 'Death, the Environment, and Theology', *Dialog* 57, no. 4 (December 2018), pp. 287–94, https://doi.org/10.1111/dial.12437.

6 Pihkala Panu, 'Anxiety and the Ecological Crisis: An Analysis of Eco-Anxiety and Climate Anxiety', *Sustainability* 12, no. 19 (23 September 2020), p. 7836, https://doi.org/10.3390/su12197836.

7 Pihkala Panu, 2020, 'Climate Grief; How We Mourn a Changing Planet', *BBC Climate*, 3 April, https://www.bbc.com/future/article/20200402-climate-grief-mourning-loss-due-to-climate-change, accessed 15.06.2022.

8 Hannah Malcolm (ed.), 2020, *Words for a Dying World: Stories of Grief and Courage from the Global Church*, London: SCM Press.

9 Ashlee Cunsolo and Karen Landman (eds), 2017, *Mourning Nature: Hope at the Heart of Ecological Loss and Grief*, Chicago, IL: McGill-Queen's University Press; Ruth Valerio, 2020, 'Ruth Valerio: Climate Grief, Perseverance and Impatience', *The Hopeful Activists Podcast*, 10 January, https://gaana.com/song/ruth-valerio-climate-grief-perseverance-and-impatience, accessed 15.06.2022; Timothy Harvie, 'Political Lament: Extinction, Grief, and Embodied Silence', *Studies in Religion/Sciences Religieuses* 50, no. 3 (September 2021), pp. 419–29, https://doi.org/10.1177/00084298209960105; Lisa H. Sideris, 'Grave Reminders: Grief and Vulnerability in the Anthropocene', *Religions* 11, no. 6 (16 June 2020), p. 293, https://doi.org/10.3390/rel11060293; Philip Jenkins, 2021, *Climate, Catastrophe, and Faith: How Changes in Climate Drive Religious Upheaval*, New York: Oxford University Press.

10 Another fascinating point of discussion, and perhaps worthy of further theological exploration by those who write about climate grief, is that anxiety and grief within environmental engagement can also be a way of avoiding the painfulness of an eventual ecological threat. I would argue that this response too is one in which there is privilege. The above exploration is well mapped out in Renee Lertzman, 2015, *Environmental Melancholia: Psychoanalytic Dimensions of Engagement*, London: Routledge.

11 Farhana Sultana, 'Critical Climate Justice', *The Geographical Journal* 188, no. 1 (March 2022), pp. 118–24, https://doi.org/10.1111/geoj.12417.

12 Sayantani Ghosh, 2020, 'Covid-19: Grief Is a Thing of Privilege', *Feminism India*, 27 April, https://feminisminindia.com/2020/04/27/covid-19-grief-thing-privilege/, accessed 15.06.2022.

13 Nylah Burton, 2020, 'People of Colour Experience Climate Grief More Deeply than White People', *Vice*, 14 May, https://www.vice.com/en/article/v7ggqx/people-of-color-experience-climate-grief-more-deeply-than-white-people, accessed 15.06.2022. As I type this sentence, I am reminded of Audre Lorde's incredible words in her poem 'A Litany for Survival'. See Audre Lorde, 2000, *The Collected Poems of Audre Lorde*, New York: Norton.

14 Shelby Brewster, 2021, '"Grieving Well": On Mourning, Extinction, and White Privilege', *Environmental History Now*, 19 March, https://envhistnow.com/2021/03/19/grieving-well-mourning-extinction-and-white-privilege/, accessed 15.06.2022.

15 Sarah Jaquette Ray, 2013, *The Ecological Other: Environmental Exclusion in American Culture*, Tucson, AZ: University of Arizona Press.

16 Sarah Jaquette Ray, 2021, 'Climate Anxiety Is an Overwhelmingly White Phenomenon'. *Scientific American*, 21 March, https://www.scientificamerican.com/article/the-unbearable-whiteness-of-climate-anxiety/, accessed 15.06.2022; Harsha Walia, Robin D. G. Kelley and Nick Estes, 2021, *Border & Rule: Global Migration, Capitalism, and the Rise of Racist Nationalism*, Chicago, IL: Haymarket Books.

17 Ray, *The Ecological Other*. This is also a topic studied by: Garland E. Allen, '"Culling the Herd": Eugenics and the Conservation Movement in the United States, 1900–1940', *Journal of the History of Biology* 46, no. 1 (2013), pp. 31–72; Marius Turda, 2014, *Eugenics and Nation in Early 20th Century Hungary*, Basingstoke: Palgrave Macmillan; Bikrum Singh Gill, 'A World in Reverse: The Political Ecology of Racial Capitalism', *Politics* (12 March 2021), https://doi.org/10.1177/0263395721994439, as well as the various notations made on the subject of green nationalism in Chapter 4.

18 Beverly Wildung Harrison, 'The Power of Anger in the Work of Love: Christian Ethics for Women and Other Strangers', *Union Seminary Quarterly Review* 36, no. Suppl. (1981), pp. 41–57.

19 Gregory L. Bock and Court D. Lewis (eds), 2021, *Righteous Indignation: Christian Philosophical and Theological Perspectives on Anger*, Lanham, MD: Lexington Books/Fortress Academic.

20 Amy Allen, 2021, 'Good Anger', *Political Theology Network* (blog), 2 August, https://politicaltheology.com/good-anger/, accessed 15.06.2021.

21 Vincent Lloyd, 'Anger: A Secularized Theological Concept', *Political Theology* 22, no. 7 (3 October 2021), pp. 584–96, https://doi.org/10.1080/1462317X.2021.1986200.

22 Alfred T. Hennelly, 1995, *Liberation Theologies: The Global Pursuit of Justice*, Mystic, CT: Twenty-Third Publications.

23 Ecumenical Association of Third World Theologians, Women's

Commission, Virginia Fabella and Mercy Amba Oduyoye, 2006, *With Passion and Compassion: Third World Women Doing Theology: Reflections from the Women's Commission of the Ecumenical Association of Third World Theologians*, Eugene, OR: Wipf and Stock.

24 Joshua Samuel, 'Raging for Liberative Reconciliation: Prophetic Anger in Dalit and Black Theologies', *International Journal of Asian Christianity* 4, no. 2 (27 August 2021), pp. 209–22, https://doi.org/10.1163/25424246-04020004.

25 Faustina, 1997, 'From Exile to Exodus: Christian Dalit Women and the Role of Religion', in V. Devasahayam (ed.), *Frontiers of Dalit Theology*, Delhi: ISPCK, pp. 96–7.

26 Gustavo Gutiérrez, 1987, *On Job: God-Talk and the Suffering of the Innocent*, Maryknoll, NY: Orbis Books.

27 Pope Francis, *Querida Amazonia: Post-Synodal Exhortation to the People of God and to All Persons of Good Will*, Vatican: Libreria Editrice Vaticana, §15–16, https://www.vatican.va/content/francesco/en/apost_exhortations/documents/papa-francesco_esortazione-ap_20200202_querida-amazonia.html, accessed 15.06.2022.

28 Wonchul Shin, 'Reimagining Anger in Christian Traditions: Anger as a Moral Virtue for the Flourishing of the Oppressed in Political Resistance', *Religions* (14 May 2020), https://www.mdpi.com/2077-1444/11/5/244/pdf.

29 Rosemary Radford Ruether, 'Anger and Liberating Grace', *The Living Pulpit* 2 no. 4 (October/December 1993), pp. 6–7.

30 Corey B. Walker, 2018, 'Love and a Theology of Blackness: A Meditation on James H. Cone', *Black Perspectives*, 5 September, https://www.aaihs.org/love-and-a-theology-of-blackness-a-meditation-on-james-h-cone/, accessed 15.06.2022.

31 Joshua Lazard, 2015, 'A Theology of Anger: Forgiveness for White Supremacy Derails Action and Alienates Young Black Activists', *Religion Dispatches*, 6 July, https://religiondispatches.org/a-theology-of-anger-forgiveness-for-white-supremacy-derails-action-and-alienates-young-black-activists/, accessed 15.06.2022.

32 Xolani Kacela, 'Towards a More Liberating Black Liberation Theology', *Black Theology* 3, no. 2 (July 2005), pp. 200–14, https://doi.org/10.1558/blth.3.2.200.65727.

33 James H. Cone and Cornel West, 2018, *Black Theology and Black Power*, Maryknoll, NY: Orbis Books.

34 Johnny Bernard Hill, 2013, *Prophetic Rage: A Postcolonial Theology of Liberation*, Prophetic Christianity, Grand Rapids, MI: Eerdmans.

35 Hill, *Prophetic Rage*, p. 14.

36 C. L. Nash, 'A Black Woman's Prophetic Rage: Religious Epistemology as Needed Boundary Crossing', Online, 2021.

37 Felix Wilfred, 'Listening to the World: Prophetic Anger and Sapiential Compassion', *Buddhist-Christian Studies* 34 (2014), pp. 63–6.

38 Tiffany Tsantsoulas, 'Anger, Fragility, and the Formation of Resistant Feminist Space', *The Journal of Speculative Philosophy* 34, no. 3 (24 July 2020), pp. 367–77, https://doi.org/10.5325/jspecphil.34.3.0367.

39 Maria Lugones, 2003, *Pilgrimages/Peregrinajes: Theorizing Coalition against Multiple Oppressions*, Feminist Constructions, Lanham, MD: Rowman & Littlefield; also Jen McWeeny, 'Liberating Anger, Embodying Knowledge: A Comparative Study of María Lugones and Zen Master Hakuin', *Hypatia* 25, no. 2 (2010), pp. 295–315.

40 Lloyd, 'Anger: A Secularized Concept'.

41 Natalia Imperiatori-Lee, 2019, 'Review: When Wrath Is Not a Sin', *America Magazine*, 13 May, https://www.americamagazine.org/arts-culture/2019/05/02/review-when-wrath-not-sin, accessed 15.06.2022.

42 Myisha V. Cherry, 2021, *The Case for Rage: Why Anger Is Essential to Anti-Racist Struggle*, New York: Oxford University Press.

43 Audre Lorde, 1981, 'The Uses of Anger', Keynote Address, CUNY.

Index

Anthropocene 103, 114–15n44
anti-colonial resistance, movements 7, 8, 49, 59, 123
Asian context 44, 49–50
Asian Theology 4, 48–9

Bandung conference 43–5, 105
Bhambra, Gurminder 68, 73
Black Theology 43, 125
Boff, Leonardo 13, 14, 41, 51–4
Buddhist 97, 100, 105, 126

Canada 2, 3, 13, 19, 57, 84, 104, 105
Caritas 3, 19–20, 31, 96, 105
Chipko movement 18, 99
Christian Aid 71, 96
colonialism 15n15, 27, 43, 49, 57, 60–1, 71, 73, 127–8
Cone, James 13, 41, 53, 54, 56, 125

Dalit 43, 69, 124
Deane-Drummond, Celia 18, 20–2, 55
deep incarnation 18, 55
Deloria, Vine 60
Divine Love 124
dukha 26, 29
Dutch East India Company (VOC) 75, 76, 77, 89n47

ecological conversion 3, 12–13, 16–27, 30, 33, 40–41, 52–3, 118, 123, 128
Ecumenical Association of Third World Theologians (EATWOT) 44–8, 107
empire 5, 59, 71, 73–4, 77, 125
Eucharist, eucharistic 20, 22–3, 26, 50
exploitation 6, 8, 25, 27, 32, 41, 49, 61, 68, 82, 83, 100, 109, 118, 120, 124
Extinction Rebellion (XR) 5, 9, 100–101

Fanon, Frantz 66, 95, 103

Feminism, feminist 32–3, 43, 52, 55, 69, 83, 106, 107–8, 125
ecofeminism, ecofeminist 33, 88n40
Filipino (Bishops' Conference) 12, 17–19, 21, 30, 41, 96
Francis of Assisi 23, 25, 28
Francis (Pope) 20–9, 30–3, 50, 52, 54, 60, 124
Gebara, Ivone 32, 41, 45, 46, 52, 53
gender, genders, gendered 8, 13, 24, 31, 32, 46, 50, 54, 105, 126
Ghosh, Amitav 76
Gnanadason, Aruna 41, 53, 55, 84
grassroots activism 97
grief 2, 4, 7, 9, 12, 13, 14, 119–22, 125, 126–8
Gutiérrez, Gustavo 42, 69, 124

India 16, 18, 73, 75, 76, 83, 99, 121
indigenous communities 6, 13, 51, 57–8, 60, 79, 80, 98–9, 104, 121
indigenous theology 41, 57–8, 59, 68, 69, 98
Indonesia 50, 75, 76, 77, 78
injustice 4, 6, 13, 22, 26, 27, 31, 68, 69, 83, 109, 124
 epistemic 68, 69
 structural 42

integral ecology 20, 24, 25, 28, 29, 31, 54
Intergovernmental Panel on Climate Change (IPCC) 16

John Paul II 17, 21–2
justice 4, 9–10, 11, 14, 23, 24, 25, 26, 28, 40, 41, 43, 49, 53, 57, 67, 68, 70, 93, 95, 103, 104, 105, 106, 107, 118, 122, 123, 124, 125, 127, 128
 climate 7, 19, 21, 22, 32, 96, 119
 ecological 3–6, 9–10, 13, 26, 32, 33, 48, 51–4, 67, 70, 84, 95, 96, 100, 105–6, 109, 122, 126–8
 environmental 5, 94, 100, 103, 105
 gender 32
 global 57, 69, 95
 racial 4, 53–4, 119
 radical 9, 122
 reparative 100
 social 3, 30, 40, 52, 96, 98, 105, 107
 transformative 7
justice narrative 3, 4, 9–11, 13, 14, 128

Katoppo, Marianne 50
King Jr, Martin Luther 125

Latour, Bruno 103
Laudato Si' 3, 8, 10, 12, 19, 20, 22, 23, 24, 25, 26, 30,

INDEX

32, 33, 50, 54, 94, 100, 109
Laudato Si' Institute 31
Liberation theology 4, 12–13, 17, 25, 30–1, 33–4, 40–2, 51–2, 55, 57, 106, 123–4
Lorde, Audre 127

Malthusian 103
Minjung theology 43
mobilization 9, 47–8, 50, 72, 81, 82, 84, 94–5, 96, 97, 98, 101, 103–4, 108, 111n13, 113n35, 120–2, 125

Neo-colonialism 10, 128
Non-Aligned Movement 45

Patriarch Bartholomew 12, 21, 26, 29
patriarchy 32, 45, 46, 53
plantation 83
Perry, Keston 6
Pieris, Aloysius 47, 49
poor, poorest, the poor 5, 6, 13, 7, 12, 13, 19, 20, 25, 33, 42, 46–8, 51, 52, 106, 107, 109
poverty 13, 24, 31, 41, 42, 29, 51, 52, 105, 109
prophetic 19, 26, 30, 67, 95, 109, 121, 123, 125–7

Querida Amazonia 12, 25–7, 30

race 6, 13, 41, 45, 48, 71, 74
racialization 71, 73
racialized 2, 5, 7, 8, 9, 14, 77
racism 6, 7, 53–4, 67
rage 4–6, 9–12, 14, 20, 27, 33, 41, 49, 108, 118–28
   generative 4, 7, 26, 43, 123
   prophetic 26, 30, 67, 95, 122–3, 126, 127
Prophetic Rage: A Postcolonial Theology of Rage (Johnny Bernard Hill) 125
reconciliation 12, 128
repentance 7, 12, 22, 30
resistance 8, 12, 18, 49, 51, 57–9, 60, 68, 123, 127

Said, Edward 68, 73
Sealey-Huggins, Leon 6, 52
silencing 6, 7, 9, 10, 68
sinfulness 12, 19, 28–30, 43, 51–2, 67, 84
   ecological 3–4, 12, 13, 16, 19–20, 27–30, 40–1, 55, 60, 67, 118, 128
Sri Lanka 1–3, 13, 46, 57, 67, 75–6, 79, 80–2, 84, 94, 105–7

Third World 4, 40–1, 43, 44–9, 52, 60, 66, 69, 107, 109, 124
Tilley, Lisa 75

womanist 43, 53, 56, 126
Wynter, Sylvia 74, 103

www.ingramcontent.com/pod-product-compliance
Lightning Source LLC
Chambersburg PA
CBHW072008290426
44109CB00018B/2176